THE DEVELOPMENT OF INK, PAPER AND PRINTING IN ASIA

"Ink is the great missive weapon
In all battles of the learned…"

Jonathan Swift.

THE DEVELOPMENT OF
INK, PAPER AND PRINTING IN ASIA

BERTHOLD LAUFER

ORCHID PRESS

THE DEVELOPMENT OF INK, PAPER AND PRINTING IN ASIA
Berthold Laufer (1874-1934)

Chapters I to IV were first published in *Printing Ink, a History*, Harper and Brothers, New York, 1927; Chapter V was previously published in a private edition of 250 copies for the Caxton Club, on the occasion of a lecture presented by the author to that group, in Chicago, 1931.

First edition as presently compiled and edited,
copyright © Orchid Press, 2019.

ORCHID PRESS
P. O. Box 19,
Yuttitham Post Office,
Bangkok 10907, Thailand
www.orchidbooks.com

Protected by copyright under the terms of the International Copyright Union: all rights reserved. Except for fair use in book reviews, no part of this publication may be reproduced in any form or by any means, electronic or mechanical, including photocopying, recording, or by any information storage or retrieval system without prior permission in writing from the copyright holders.

Front cover illustration: Section of a Korean wooden printing block, for pages of 'The Collected Works of Eun Wha,' Yi Dynasty, 17th or 18th century. Private collection.

Back cover image: Chinese ink with image of Guanyin, product of the Hu Kai-wen studio, late Qing Dynasty, China. Private collection.

ISBN 978-974-524-110-7

CONTENTS

	Page
Editor's Notes	vii
I. The History of Ink in China and Korea	1
II. The History of Ink in Japan	56
III. The History of Ink in Central Asia	60
IV. The History of Ink in India	64
V. Paper and Printing in China and Korea	72
Index	101
About the Author	105

ILLUSTRATIONS

Fig. 1. Frontispiece of the *Vajracchedikā*	23
Fig. 2. Chinese ink makers straining oil.	29
Fig. 3. Compounding Chinese ink.	31
Fig. 4. Ink-cake design from the *Fang shi mo p'u*.	41
Fig. 5. Ink-cake design from the *Ch'eng shi mo yüan*.	43
Fig. 6. Rubbing of an ink-stone in the form of a lute, Song dynasty.	53
Fig. 7. Bamboo papermaking tools.	81
Fig. 8. Collecting bamboo pulp on top of a screen.	83
Fig. 9. Cutting movable Chinese type.	94
Fig. 10. Setting movable Chinese type.	95
Fig. 11. Sample page of book printed in Korean *Chongnija* movable type.	99

EDITOR'S NOTES

Berthold Laufer's many publications on Chinese culture primarily employed the romanization system for Chinese originally developed by the École Française d'Extrême Orient (EFEO); later his writings occasionally used the more widely-known Wade-Giles (W-G) romanization.

The EFEO system of Chinese romanization, which shares some characteristics with W-G, was developed in the early 20th century and designed not specifically to represent Standard Mandarin but rather to address a wider mean of the sounds of the various Chinese regional dialects. This was a natural fit for Laufer, a polyglot who mastered over a dozen languages, and worked not only in Mandarin but also in several other Chinese dialects, as well as archaic Chinese, Manchu, Mongolian, Tibetan, Japanese, Sanskrit, Pali and other Asian and European languages.

In editing this material, we have chosen to avoid modification of the text as far as possible, leaving Laufer's choice of romanization mostly as first written. We have, however, changed some widely used proper nouns into their more common W-G forms when these were encountered in EFEO romanization in the text—for example, K'ien-lung to Ch'ien-lung [Py. Qianlong] in the case of one of the most renowned emperors of the Ch'ing [Qing] dynasty. We have also deleted hyphens in many of the place-names that Laufer quotes—for example, Su-chou to Suchou [Py. Suzhou], etc.

As for other additions to the original work, all illustrations aside from Figures 2 and 3—which were included in *Printing Ink, a History*—have been added as a complement to Laufer's text by the present publishers. The occasional footnote has also been added to update the author's historical statements, when, for example, important new evidence has become available since Laufer's work was first published.

CHAPTER I

THE HISTORY OF INK IN CHINA AND KOREA

THE celebrated calligrapher Wang Hi-chi (AD 321-379), whose handwriting is said to have been "light as floating clouds and vigorous as a startled dragon," is credited with the dictum, "Paper represents the troops arrayed for battle; the writing-brush, sword and shield; ink represents the soldier's armor; the ink-stone, a city's wall and moat; while the sentiments of the heart symbolize the chief commander." In this saying the mental attitude of the Chinese toward the arsenal of the learned is well crystallized: paper, brush, ink, and ink-slab are the four great emblems of scholarship and culture, inventions which the Chinese may justly claim as their own, which constitute the fundamentals of their civilization, and which have largely contributed to make them a nation of studious, cultured people.

In extolling the art of printing as one of the great achievements which has remodeled our intellectual life, we must not overlook the fact that the merit of this invention rests to a lesser degree on the basic idea than on its primary conditions—the existence of an economic material suitable for writing and printing and easy to manufacture in large quantity, and a medium that will permanently fix the written thought to the paper. That rag-paper is a Chinese invention and that the Arabs transmitted the method of its manufacture to Europe is a fact established beyond any doubt, not only through historical records, but also through archaeological discoveries and microscopical and chemical analyses of ancient paper remains.

The same cannot be said about ink: Egypt, the ancients, and mediaeval Europe were familiar with ink of different kinds; but the Chinese product is so superior to anything accomplished in the West that for centuries it was employed by artists of Europe under the misnomer "India ink" and is still unrivalled.

The date for the first manufacture in China of ink in the proper sense of the word is variously given in current literature. Palladius, a prominent Russian Sinologue, writes that ink is said to have first been produced in AD 220; and Geerts (*Les produits de la nature japonaise et chinoise*, 1878, p. 197), in accordance with S. Julien, gives the date more specifically as that of the Wei and Tsin dynasties (AD 220-419). This indeed is the period commonly fixed in Chinese sources. In consequence of a misprint in M. Jametel's little book *L'Encre de chine* (1882, p. xi), where the date of the Wei is given as "220 à 260 [instead of 265] avant [instead of après] J.-C.," several authors have adopted the error in assigning the invention to the third century BC. Thus F.M. Feldhaus (*Technik der Vorzeit*, 1914, col. 1198) and Rein (*Industries of Japan*, p. 417) even turn the figures around, giving 260-220 BC as the date for the invention of ink. Giles (*Glossary of Reference on Subjects connected with the Far East*, p.132) states that ink was used all over the empire since the third century of our era, though, according to one native authority, it was manufactured as early as 140 BC.

While Chinese records give us a name for the inventor of rag-paper and the writing-brush, there is no name on record for the inventor of ink, simply for the reason that ink is not the invention of an individual. The situation is the same as that with regard to porcelain: Porcelain is not an invention that can be attributed to the efforts of an individual; but it was a slow and gradual process of finding, groping, and experimenting, the outcome of the united exertions of several centuries and generations. The same observation holds good for the history of ink. It took the Chinese several centuries of tests and trials until they eventually discovered an acceptable formula for a good ink, and even after this discovery they made constant improvements and developed the method, as they also enlisted new materials. It is one of the outstanding examples of progress in Chinese technology and an eloquent refutation of the dogma of the stationary character of Chinese culture.

If, in accordance with Chinese conception, ink properly so-called was only the result of the labors of the early middle ages (third to the beginning of the fifth century AD), our historical inquiry is mainly concerned with three questions: (1) What were the writing-materials in the times of the earliest antiquity of China? (2) What was the medium of writing in the age of the closing antiquity (period of the two Han dynasties, 209 BC-AD 220), when the writing-brush and finally paper (from AD 105) existed? (3) What did the invention of the Wei and Tsin periods consist of, and how was it further developed?

In a certain class of popular Chinese books of recent date whose main object it is to trace the history of cultural objects and inventions, and which have the undisguised tendency to advance them as far as possible into the dim past, it is boldly asserted that the beginnings of ink and ink-slabs go back into the mythical days of the emperor Huang-ti (alleged date 2698 BC), and some even give as the name of the "inventor" Tien Chen, supposed to have lived at that time. No such tradition exists in any ancient book as the *Chu shu ki nien* ('Annals Written on Bamboo Tablets') or the *Shi ki* ('Historical Memoirs') of Se-ma T'sien. This modern contruction in purely fictitious and arbitrary, and is contradictory to all historical facts known in the case.

In prehistoric China knotted cords were used to convey messages, primarily for government business. The Tibetans and certain aboriginal tribes in south China shared a similar tradition. In early historic times contracts were made by means of wooden tallies in which notches were carved with a knife; the creditor, for instance, received the left; the debtor, the right half of the tally. Under the Shang dynasty (1783-1123 BC) bone and tortoise-shell served as the conveyance of writing, the characters being slightly incised in the surface; such bones were chiefly inscribed for purposes of divination, and many have been unearthed during the last two decades. The earliest form of Chinese script is preserved on them. Further, we have from the early dynasties inscriptions on bronze vases and bells, the writing being produced in the wax mould, and being either incised or raised. Tablets of jade were used for writing by the emperor; tablets of ivory, by the nobles and higher officials. The most common material, however, particularly under the Chou dynasty (1122-247 BC) consisted of bamboo slips or square wooden splints which were perforated at their upper ends and fastened together by means of a silk cord or fine leather strip. The main difference between the utilization of bamboo and wood was this, that a message containing upwards of a hundred words was written on bamboo slips; when it contained less than a hundred words, on wooden boards. The bamboo tablets were naturally narrow, and could be piled up in any required number formed into a pack. The wooden documents, being too heavy to allow of a combination of many, served only for brief texts, as official acts and regulations, statistics of the population, and prayers, but they could not be united into books.

Early literature was handed down on bamboo slips of different lengths, each slip usually containing a single line of writing varying from eight to twenty-five or thirty words, and inscribed on one side only. Such books, of course, were exposed to many causes of destruction, chiefly from humidity and pernicious insects, so that bamboo books of early antiquity have long since disappeared. Another inconvenience of these books was their heavy weight.

Neither writing-brush nor ink was invented in those early days, and the bamboo and wooden memoranda were inscribed by means of a pointed bamboo or wooden stylus (*pi*) dipped in a black varnish (*ts'i*). The bamboo or wooden stylus has survived in Tibet and among several other tribes akin to the Chinese. Varnish, according to the unanimous opinion of all competent scholars of China, was the earliest vehicle of committing thoughts to writing. What the composition and preparation of this ancient varnish was, however, is not known; but it is more than probable that it was a product obtained from the sap of the lacquer or varnish tree (*Rhus vernicifera* D.C., family *Anacardiacea*), a sumach indigenous to China. In fact, this tree is designated by the same word *ts'i* as applied to the varnish used in writing. The corresponding Tibetan word *r-tsi* denotes any thick vegetable sap, varnish, or paint. The tree was cultivated by the Chinese in ancient times; in the Book of Songs (*Shi king*) it is mentioned three times, in one case as having been planted by Duke Wan. In superior quality its varnish was produced in Yuchou in the province of Honan and in Yenchou in the western part of the province of Shantung, as may be gleaned from the chapter 'Yu-kung' inserted in the *Shu king* ('Book of Historical Records').

In the *Chou li*, the State Handbook of the Chou dynasty (1122-247 BC), varnish is referred to as being applied to bows, to spikes of chariot-wheels, and to hides used for drums. The fact that it was regarded as a precious substance becomes evident from the Book of Rites (*Li ki*), which says that it was employed for coating the covers of the coffins of princes and the highest officials, but that this privilege was not conceded to plain officials. When a prince ascended the throne, his coffin was made and stored away, a coat of varnish being laid on once a year. The Chinese character for the tree consists of the symbols for wood and water written one above the other, alluding to the sap oozing out and dripping down the trunk. The varnish is extracted by making a horizontal slit upon the tree, and this can be done throughout the warm season, from April to the end of October. The varnish released in the spring is least valuable, because it is very watery. The autumn product is much thicker, but also granulous and slow in exudation. Midsummer is the best time for the harvest, and the varnish is then at its best as to quality and quantity. The tapping, as a rule, begins when the tree is from nine to ten years old. The tree was introduced from China to Japan (Japanese *urushi-no-ki*), where it is eagerly cultivated for the same purposes as in its mother country—the preparation of a lacquer from the sap.

There is direct testimony for the fact that books written on bamboo or wooden tablets by means of varnish were actually produced. In AD 279 a tomb was opened at Ki in the prefecture of Wei-hui, province of Honan, and yielded several ancient manuscripts of which we have a contemporaneous record: it is stated that they were written with varnish in "tadpole" characters, an ancient form of script. One the other hand, it is reported that one of the manuscripts discovered in the tomb of Ki, and interred there in 299 BC, the famous *Mu t'ien tse chwan* (the romantic narrative dealing dealing with King Mu's travels to the west) was written on bamboo slips, each containing forty words, with ink (*mo*); and this is the word still used for "ink." This word appears frequently during the Han period (209 BC-AD 220), and even a few centuries earlier.

Chou Sho, the councilor of Ch'ao Kien-tse, who died in 458 BC, said to his master, "with my brush [*pi*] soaked in ink [*mo*] and the tablet held in my hand I shall watch over the faults of your highness." In the Annals of the Later Han Dynasty we read of large and small pieces of ink, and even the term "paper and ink" (*chi mo*) occurs. In view of this fact it is curious that the majority of Chinese scholars who have made ink the subject of special research are agreed on the point that ink in the modern sense was made but as late as the age of the Wei and Tsin dynasties; that is, the period from AD 220-419. In order to understand and reconcile these anomalies it is necessary to scrutinize the subject at closer range.

The word used throughout the centuries and still at the present time for the designation of writing and printing ink is *mo* (ancient form *ma-g*, Canton *mok*, Amoy *bat*, Yunnan *muk*, Korean *mik*, Japanese *moku* or *boku*).The question arises, what notion was conveyed by this word in times prior to the invention of real ink? The etymology of the word gives no direct response to this query. Its primeval and fundamental significance is "black, dark, obscure"; and in this sense it is used in a passage occurring in the work of the philosopher Mongtse (called Mencius) in the fifth century BC, where the face of a minister is described as deep black (*shon mo*). From a comparative viewpoint ancient Chinese *mag* corresponds to Tibetan *nag* ("black"), *s-mag* ("dark"), and *s-nag* ("ink"), which goes to show that both in Chinese and Tibetan the word "black" has assumed the meaning "ink," as soon as the invention of ink was made; in fact, the word *nag* or *mag* (thus in some Karen dialects) for "black" is common to all Indo-Chinese or Sinic languages.

In consulting the written symbol or character for the Chinese word *mo*, we find that it is not a simple, spontaneous formation, but presents a composition of two well-known signs, those for "black" (*hei*, anciently *gag*) and "earth" (*t'u*, anciently *du*). The meaning connoted by the character, accordingly, is "black earth" or "black clay," and may hint at the fact that a mineral or clayish substance of black color may be hidden under this term. This opinion is indeed advanced by Chinese writers. Thus Li Shi-chen, author of the famous herbal *Pen ts'ao kang mu* in the latter part of the sixteenth century, infers from the formation of the character for *mo* that the ink of the ancients was made from black earth (*hei t'u*), referring also to the definition in the ancient dictionary *Shwo wen* (about AD 100) that *mo* is a kind of earth formed by smoke and glue. It must be borne in mind, however, that any such conclusions as to material, merely deduced from the construction of written symbols, are usually fraught with danger or may be deceptive, that the present explanation is more or less an afterthought, and that the old contemporaneous tradition is lost.

Several passages in ancient texts show us that *mo* was a black pigment. The philosopher Mong-tse, mentioned above, speaks in another chapter of his work of the carpenter's marking-line, expressing it by the compound *sheng mo* ("string and ink," or more cautiously "string and black substance *mo*"). It is clear that the carpenters of his time, in the same manner as those of the present day, must have availed themselves of an instrument for marking lines in black.

Another instance of the ancient application of the word *mo* occurs in the penal code of the Chou dynasty first issued on bamboo slips in 501 BC in which five kinds of punishment are laid down, the first of these being branding of a criminal's forehead. This process is denoted by the term *mo*, which in this case is imbued with verbal force ("to blacken, to brand").

We are not informed as to what this black pigment used for the mark of infamy was; it was applied to the forehead by means of an incision with a cutting instrument, and must have been of a rather indelible nature (cf. Couvreur, *Chou king*, p. 386).

Among the ancient Chinese the tortoise was one of the principal vehicles of divination. The carapace of the animal was coated with a layer of a black pigment (*mo*) and exposed to a fire. Thereupon the delineations of the cracks produced by the action of the fire were examined, and the will of Heaven was read from them. This process was styled *ting mo* ("to determine the pigment"; Couvreur, *Li Ki*, Vol. 1, p. 682). It was the rule, however, to burn the tortoise-shell without the application of a pigment.

The invention of the writing-brush must have acted as a stimulus to the improvement of all writing materials. Traditionally, the brush is credited to Mung T'ien, who served as general to the first emperor Ts'in Shi, and who died in 209 BC. This tradition, however, is to be taken *cum grano salis*. It is not recorded by a serious historian like Se-ma T'sien, whose *Shi ki* is the chief source for the history of that emperor's reign; and some Chinese authors are inclined to think that the invention of the brush was merely attributed to Mung T'ien for the glorification of his imperial master, who wished everything to begin from his reign. It would be absurd to assume that the general was the first Chinese who ever invented a brush; so simple an implement must doubtless have existed centuries before his time.

The eminent scholar Yen Shi-ku (AD 579-645) probably hits the truth with the following comment: "The tubes of the ancient writing-brushes were made from dried wood with deer's hair backed by sheep wool, but there were no bamboo tubes with hare's hair; this was the work of Mung

T'ien." In other words, Mung T'ien may have applied two improvements to the writing-brush, which itself pre-existed; he made it lighter in weight, with finer hair, and perhaps more elegant in form. If he really had anything to do with the whole affair, it is striking that an instrument exalted by the literati and essentially one of the learned owed its perfection to a man of the much despised military class; or, we might rather say, it is surprising that popular imagination, in seeking for the inventor of the writing-brush, has fastened the honor on an old general who is not known as a writer.

Another important innovation took place in the third century BC, presumably under the Ts'in dynasty (246-207 BC), when a paper made from silk refuse including both raw and woven silk came into being. This silken paper was preceded by bands of silk stuff serving as writing material. The refuse from silkworm cocoons was soaked and beaten in water in order to eliminate coarse particles. This mass was reduced to a paste, and thus purified was spread over a fine bamboo mat mounted on a wooden frame. This served as the mould on which the paste precipitated, and when dried, produced a sheet of paper. Such bamboo mats are still utilized in the modern manufacture of paper from bamboo or tree-bast fibres (see below Ch. V, Figs 7 and 8).

It is clear also that the underlying principle of the subsequent manufacture of rag-paper was forestalled by that of silk paper. Since the writing-brush was perfected in the same period, we can hardly call this coincidence accidental, but must admit that the two inventions were dependent one on the other; hence the further conclusion is justified that the two again may have stimulated the production of a better expedient for writing, which resulted in the black pigment called *mo*. At the outset it is not very likely that varnish continued for silk material and silk paper, nor is it likely that the ancient wooden stylus was applied to silk. Thus there are good technical reasons for the conviction that the varnish of early antiquity was gradually replaced by a more convenient substance in the three or four centuries preceding our era.

A still more powerful impetus to the improvement of writing-ink was received from the invention of bark and rag paper by Ts'ai Lun,[1] a native of Kweichou Province, in AD 105. Ts'ai Lun considered that both silk and bamboo tablets were inconvenient writing-materials, the former being too expensive, the latter too cumbersome and perishable. Hence he conceived the idea of using bast-fiber, hemp, and old rags like fishing-nets for making three kinds of paper.

This invention created a profound impression on the contemporaries, who at once turned it to practical use. Although Ts'ai Lun perpetuated and advanced a pre-existing process, and his principal merit consisted in the substitution of little valuable or even valueless substances, which simultaneously yielded better results, for the comparatively costly silk-refuse, he must be honored as the man to whom we are indebted for one of the most far-reaching discoveries ever made in the annals of technology. Without paper there would have been no adequate record of the past, no progress, no science; it marks the dawn of civilization, it sets off civilization from savagery.

Thus the life of the Han dynasty, during the last century of its existence, was signally enriched by the acquisition of paper. Nevertheless, in the outlying colonial possessions, the use of wooden tablets persisted with conservative force. A great number of these were rescued from the sands of Chinese Turkestan by Sir Aurel M. Stein, and have been edited and translated by E. Chavannes (*Les Documents chinois découverts*

---

[1] Further details on the life of Ts'ai Lun may be found in Ch. V, p. 78-80.

dans les sables du Turkestan oriental, Oxford, 1913). The wooden slips studied by Chavannes range in date from 98 BC to AD 153. He observes that a goodly number of these are inscribed with characters of extreme finesse, and that their delicate traits could have but resulted from the use of a brush; unfortunately he is silent as to the ink. In the reproductions of the documents the writing appears in black ink, in appearance not different from the Chinese ink to which we are accustomed. This, of course, does not mean that the Han ink might not have been an entirely distinct affair. Microscopical and chemical analysis would be the only means of solving the problem, and it hoped that the authorities of the British Museum will consent to having this investigation made some day.

Meanwhile, we are thrown back on Chinese literary sources. It is certain that in the Han period an ink-like substance for writing, called *mo*, was utilized; but what its composition was, is not revealed. No recipe has been handed down from that epoch. The fact that it was different from the later ink is manifest from the persistent tradition that this product made its appearance only under the Wei and Tsin. Ch'ao Shwo-chi of the Sung period, who wrote an interesting treatise on the technology of ink, opens it by saying that formerly two kinds of ink were in use—one prepared from pine-tree lampblack (*sung yen*), another styled "mineral ink" (*shi mo*, litrally "stone ink"). The latter, he comments, has disappeared since the Tsin and Wei periods, but he does not explain what it was. He assumes that under the Han period ink was also manufactured from pine charcoal, as *yu-mi mo* ("ink from *yu-mi*") is mentioned at that time, and the Chunnan mountains at Yumi (in Shensi Province) were covered with pines; this is a theory which remains in the realm of conjecture.

T'ao Tsung-i, who wrote the *Cho keng lu* in AD 1366 (under the Yüan dynasty), states that in times of earliest antiquity there was no ink, documents being written by means of a piece of bamboo dipped in varnish, and that in the middle period of antiquity they utilized the sap of *shi mo*, which is believed by some to be identical with the stone of Yen-ngan in Shensi (that is, rock-oil or kerosene). This mineral product is described by the Chinese as a stone grease floating on the surface of water, like varnish, and collected to be burnt in lamps or made into torches. From the Sung period onward the lampblack from petroleum was used in the manufacture of ink, called *Yen ngan shi i* ("secretion from the stone of Yen-ngan"), these words being engraved on the ink-cake.

It is on record that Lu Ki or Lu Shi-heng, who lived in the latter part of the third century AD, one day ascended the *T'ung ts'iao t'ai* ("Copper Sparrow Terrace"), a tower built by the famed Ts'ao Ts'ao in AD 210, and found several jars full of *shi mo* collected and stored by Ts'ao Ts'ao, such as existed no more at any later period. Subsequent authors indulged in speculations as to the material contained in these jars. In this case it is again assumed by some that this substance was identical with the lampblack produced from kerosene (*shi chu yen*, "stone torch smoke") in Shensi, mentioned by Shen Kwa in his *Mong k'i pi t'an*, written in the middle of the eleventh century. It is supposed also that the cosmetic used for painting the eyebrows by the women of the palace of the Sui dynasty (AD 583-617) was a kind of *shi mo*.

Under the term *shi mo* quite a number of different minerals appear to be confounded. *Shi mo* is also a synonym of *shi tan* ("mineral coal"), and Li Shi-chen (*Pen ts'ao kang mu*, Ch. 9, p. 20) affirms that in times of antiquity bituminous coal was utilized for writing. The possibility of this cannot be denied. Incidentally, one Chinese author declares that lead was anciently used for writing. Several minerals were formerly utilized as substitutes for ink. The *Yün lin shi p'u*, a treatise on economic mineralogy written by Tu Wan (pseudonym Yün-lin) in AD 1133 (Ch. 3, p. 11b), says that in Kweichou (prefecture of I-ch'ang, Hupei Province) there are black stones appearing in the water of the Yangtse River, of coarse substance, which can be ground and will yield an ink. This stone is called *ta t'o shi* ("stone of the great river"), *t'o* being the local name for the Yangtse among the inhabitants of the gorges near I-ch'ang, who prize this stone very highly. This same work mentions a stone of Ts'ingchou in Shantung, found deep in the soil, being carved and ground, not containing much ink, but used locally. A "dragon-tooth stone" (*lung ya shi*), according to the same author, is found in the district of Ning-hiang in the prefecture of Yo-chou, Hunan Province, both in water and in the mountains, of purple color, somewhat glossy, capable of being ground into ink and rather appreciated by the people of the place. Finally, in the river of the district of Fen-i in the prefecture of Yüan-chou, Kiangsi Province occurs a stone, dark in color, hard and bright, sonorous when struck, gathered by the natives in the water, and ground into ink, which is suitable for the brush; but the material is so coarse that instruments for cutting and grinding it are required. For Kwangtung Province, mineral ink mountains furnishing writing-ink of excellent quality are mentioned as early as the Tsin period in the *Kwang chou ki* of Ku Wei.

There is another kind of *shi mo*, which is identified with *hei shi chi* ("grease of black stone"), described as sticking to the tongue when licked and used for writing, as well as for painting the eyebrows (much practiced in ancient China). This is doubtless graphite (Geerts, *Produits*, p. 203). F. de Mély (*Le lapidaire chinois*, p. 256) is inclined to take it for "sulfure d'antimoine." Now the term *shi mo* in the sense of graphite occurs for the first time in the early work *Pie lu*, the foundation of which goes back to at least the Han period, and possibly even to an earlier date. This would well indicate that graphite was one of the substances enlisted as writing material in the epoch of the Han. Besides, products of mineral coal, bitumen, and rock-oil may have been utilized; the earliest ink, accordingly, was of mineral origin, in opposition to the vegetable products laid under contribution in the following period, under the Wei and Tsin (AD 220-419).

The oldest recipe for the preparation of an ink that has come down to us is contained in the *Ts'i min yao shu*, a work on practical husbandry, written by Kia Se-hie, who lived in the fifth or sixth century AD. Unfortunately this important work is handed down in mutilated form. The original was in 92 sections, part of which were lost long ago, and much additional matter has been interpolated by subsequent editors. The recipe for ink, entitled "Method of mixing ink," is apparently incomplete, since the substance from which the lampblack is derived is not even mentioned; in some places the text is enigmatic and evidently corrupt.

Ch'ao Shwo-chi, an author of the Sung period, who wrote a very interesting treatise on the manufacture of ink, quotes three passages from the recipe of Kia, but his text is different from that found in the present editions of the *T'si min yao shu*. The principal points of the formula are as follows: Good and pure lampblack is to be pounded and strained through a sieve of fine pongee, which is placed in a vat of stoneware. The object of this process is to free the lampblack of any adhering vegetable substances so that it becomes like fine sand and dust; but as it is so light in weight, great care must be exercised in preventing it from being scattered around. Five ounces of glue are required for one pound (catty) of ink, and the sap of the bark of the *ts'in-p'i* tree (*Fraxinus pubinervus*) is dissolved in the glue. This bark is green in color like water, and contains a glutinous substance which also improves the color of the ink.

The white of five hen's eggs, one ounce of cinnabar, and the same amount of musk may likewise be added, after being well strained. All these ingredients are mixed, and the paste thus obtained must be beaten in an iron mortar with a stick thirty thousand times; the more frequently it is beaten, the better the quality of the ink. In weight a cake of ink should not exceed two or three ounces, and its size should conform to this rule; that is, it must be small, not large.

However imperfect this formula may be, it leaves no doubt that it carries the directions for a real ink, and that in principle it is identical with the process still in vogue. Even though the source of the lampblack is not clearly indicated, it follows from the context that it was of vegetable origin. Li Shi-chen, the great herbalist at the end of the sixteenth century, stated expressly that the best ink in his time was made from lampblack of pine-wood with an admixture of the sap of *Fraxinus pubinervus*, boiled together with glue as well as aromatic substances. The green bark of this tree steeped in water is still utilized for obtaining a bluish indelible ink.

It is obvious that the formula divulged by Kia Se-hie did not spring up spontaneously, but that it presents the result of long experiences and experiments conducted during several generations. In fact, he had predecessors, as we see from the *Mo king* of Ch'ao Shwo-chi of the Sung period, who quotes the "ink method" of Wei Tan or Wei Chung-tsiang of the Wei dynasty (AD 220-264); this author has also the thirty thousand beatings—doubtless an exaggeration. Coming down to the T'ang dynasty, Wang Kiün-te, an ink manufacturer, who used a stone mortar, speaks more moderately of from two to three thousand beatings. Under the Ming, toward the end of the sixteenth century, a wooden mortar with a metal pestle was employed, and one was satisfied with several hundred beathings. There is every reason to believe that Wei Tang is the first originator of an ink formula, as handed down in the *Ts'i min yao shu*, and to a certain degree may be regarded as the real inventor of ink as still manufactured. His name is the first which appears in the *Mo shi*, a history of ink manufacturers by Lu Yu of the Yüan dynasty, who observes that there is no great difference between the formulas of Wei Tan and Kia Se-hie. One of the former's peculiarities was that he mixed one ounce of genuine pearls and a half ounce of musk in his ink.

In the *T'ai p'ing yü lan* published by Li Fang in AD 983 (Ch. 605, p.5) the recipe, as contained in the *Ts'i min yao shu*, is in fact quoted as that of Wei Tan, the text being substantially the same, with the one exception that is an "ounce of cinnabar" is replaced by an "ounce of genuine pearls". We must therefore admit that Kia Se-hie merely copied Wei Tan. The latter lived from AD 176 to 251, to the age of 75. His ink was still renowned at the end of the fifth century. Siao Tse-liang, a prince of the Southern T'si dynasty (about AD 484), in one of his letters, speaks admiringly of "Chung-tsiang's ink, every drop like varnish!"

Chinese literature on ink is considerable. It begins to develop under the T'ang dynasty when a treatise on ink (*Mo king*) by Ch'eng Lao-po existed. This work, however, has not survived, but a few brief extracts are preserved in the *Yün sien tsa ki*, written by Fung Chi in the commencement of the tenth century, and in later works. Several technical treatises on ink were written under the Sung, above all, the *Mo king* by Ch'ao Shwo-chi, which is reprinted in the great cyclopaedia *T'u shu tsi ch'eng*. This author devotes a systematic discussion to the various kinds of pine-trees suitable for lamp-black, with exact enumeration of the mountains and localities where they grow; then he has discourses on charcoal, glue, sifting and the processes of mixing and pounding, chemical ingredients, drying, grinding, color, sound, weight, new and old ink, preserving ink, proper season for making it, workmen and manufacturers.

Another treatise on ink, entitled *Mo p'u*, was published by Su I-kien at the close of the tenth century. The same author wrote a book on paper (*Chi p'u*), and in AD 986 summarized his experience in a comprehensive work which he called *Wen fang se p'u* ('The Four Departments of the Study'), This is a repository of information regarding the materials of the study, consisting of four parts which treat of writing-brushes, paper, ink and ink-pallets, with historical memoranda, essays, and stanzas. Ho Yüan of the Sung wrote a *Mo ki* ('Ink Memoirs') in which he deals with ink manufacturers; a long list of these from the T'ang to the Yüan dynasty is also inserted in the *Cho keng lu* by T'ao Tsung-i, published in AD 1366.

The most complete and interesting work of this class, however, is represented by the *Mo shi* ('Ink History') of Lu Yu of the Yüan dynasty, who gives a series of brief notices of about a hundred and forty manufacturers whose names had been handed down in connection with their productions from the Wei, Tsin, T'ang and Sung dynasties down to the Kin. He also notes the ink of the Koreans, the Kitans, and Turkestan, with a number of miscellaneous observations respecting ink appended.

In the beginning of the Ming dynasty, in 1398, appeared a manual of the ink manufacturer, by Shen Ki-swan, illustrated by twenty-seven woodcuts showing the various stages in the process of manufacture. The author was a manufacturer himself and professes to divulge only the tricks of his own trade, save some information communicated to him by a monk. The little work teems with technical routine and detail, but in comprehensiveness and clarity does not compare with the *Mo king* of the Sung. It was first translated into Russian by I. Goshkewich (in 'Works of the Russian

Mission of Peking', Vol. I, 1852; cf. W. Schott, *Entwurf einer Beschreibung der chines. Literatur*, p. 107), then from the Russian into German. A French translation with reproduction of the woodcuts is due to M. Jametel (Paris, 1882).

About 1637 Ma San-heng issued a *Mo chi* in which he treats of ink manufacturers with their productions and the marks that distinguish them. Kao Lien of the Ming wrote an essay, *Lun mo* ('Discourse on Ink'), and T'u Lung published the *Mo tsien*, a short work on ink, during the sixteenth century. The *Süe t'ang mo p'in* is a small treatise on ink, written by Chang Jen-hi in 1671, in which he classifies the productions of various manufacturers and points out the peculiar characteristics of the different kinds. The *Man t'ang mo p'in* is a similar record, supplementary to the preceding, written in 1685 by Sung Lao, who presents notices of thirty-four specimens of ink of the Ming dynasty, with their respective weights. Aside from such monographs there are innumerable references to ink in technological books like the *T'ien kung k'ai wu*, written by Sung Ying-sing in 1637, in the essay literature of the Sung, in cyclopaedias, and in the herbals (*Pen ts'ao*).

While the Ming and Manchu dynasties hardly added anything new to the technical side of the subject, great care was bestowed on the artistic embellishment of ink-cakes, and we have several excellent books issued by ink-manufacturers giving reproductions of the designs engraved upon their ink-cakes. These will be considered in more detail below.

After this digression, let us revert to the history of ink. The poet Ts'ao Chi (AD 192-232) has left a verse in which he says that ink is produced by the smoke of the green pine. Now he was a contemporary of Wei Tan (AD 176-251), and it is therefore a reasonable conclusion that Wei Tan was the first who used the lampblack from pine in the manufacture of ink. There is another formula handed down from the Wei period and attributed to a certain Ki *kung* (Mr. Ki) about whom nothing is known otherwise. His ink is said to have been made of two ounces of pine-black mixed with a little clove, musk, and dried varnish; this was compounded with glue, soaked in water, and heated over a fire, the entire process taking a full month; he produced inks of two colors—a purple ink by adding the root of *Lithospermum officinale* L. var. *erythrohizon* (the Chinese name of this plant means "purple herb"; it is still cultivated for the purple dye yielded by its root) and a bluish ink by using the bark of *Fraxinus pubinervus* (*ts'in p'i*). Ink was formerly put up in various forms. The Chinese language has a large number of numeratives by which objects are counted, and the numeratives vary according to different categories of things and notions. Under the Han ink pieces were usually counted as so and so many *wan* (that is, "pills, pellets, balls"); ink being formerly taken as medicine. It is very likely that it was first made into pills to be easily swallowed. This practice was continued under the Wei and Tsin, and under the T'ang also we occasionally hear of ink balls; but from that time onward this shape fell into disuse. It seems that in some out-of-the-way places ink is still made into balls; at least E. H. Parker (*Up the Yang-tse*, p. 135, Shanghai 1899) reports that at Ansi Ch'ang in Kweichou Province he saw exposed for sale balls of ink made from the soot of the *Aleurites* oil (wood oil or *t'ung* oil).

Further, ink pieces were formerly counted by *liang*, (a measure of capacity) and, curiously enough, by conch-shells (*lo*), which may indicate that ink was kept in shells. An early author, T'ao Tsuing-i, in AD 1366, asserts that in times after the Tsin there was a kind of ink called *lo-tse mo* ("shell ink"), as though shell powdered or burnt had formed an ingredient in its composition; but this notion surely rests on a misunderstanding, for the ancient texts contain nothing concerning such an ink, but what is spoken of is merely that someone, for instance, sent another "two shells of ink," which means ink of a quantity as two shells may hold, whether it was actually transmitted in shells or not.

The prismatic shape of ink seems to have come up under the T'ang: during his last exploration of Turkestan, Sir Aurel Stein (*Serindia*, Vol. 1, p. 316) discovered an octagonal prism of Chinese ink. On his previous journey he found in the stupa of Endere a "cylindrical piece of hard Chinese ink, drilled for a string at one end" (*Ancient Khotan*, Vol 1, p 438). The prismatic or cylindrical shape (so-called sticks) has persisted to this day, and this is the form of ink destined for common use. It is, further, made into small, flat, rectangular cakes, sometimes also cast into circular forms if required by the artistic subject impressed upon the cake (see Figs. 3 and 4 below).

Under the great T'ang dynasty (AD 618-906) the manufacture of ink took an unprecedented development. Several events conspired to contribute to this result. Under the T'ang sovereigns Central Asia was annexed to the empire, and this political expansion led to the predominance of Chinese civilization all over Asia. Ink was required in the outlying dominions, whether politically dependent or merely under the influence of the Chinese culture-sphere. It was eagerly demanded in Turkestan as well as in Tibet; in Annam as well as in Japan. The same epoch marks China's Augustan age in literature and painting which then reached their climax, and above all, it was the new invention of printing books by means of wooden blocks which gave a fresh impetus to further progress in the production of ink.

At the end of the sixth century, under the Sui dynasty, when printing first became known, the imperial library contained some 37,000 books; in the early part of the eighth century, being the most flourishing period of the T'ang, the number of works described in the official record of the imperial library amounted to 53,951 books, besides which there was a collection of recent authors, numbering 28,469 books. These figures will give an idea of the important function which paper and ink must then have performed in the national culture. Fortunately we now have at our disposal both manuscripts and prints of that epoch.

Woodcuts of the T'ang period have been discovered in the Cave of the Thousand Buddhas (Ts'ien Fu Tung) by Sir Aurel Stein (*Serindia*, p. 893) and Paul Pelliot. They illustrate, as Stein writes, the high stage of technique which the art of printing from wooden blocks attained comparatively soon after its first invention, and also the earliest use to which it is likely to have been put. A printed roll, dated AD 868 and containing in its 16 feet of length the complete text of a Chinese version of the *Vajracchedikā*, is the oldest specimen of printing at present known to exist,[2] and its fine frontispiece is the earliest datable woodcut. It shows Buddha seated on a lotus throne attended by a host of divine beings and monks and discoursing with his aged disciple, Subhūti.

---

[2] Since this was written an unillustrated woodblock-printed Buddhist text in Chinese characters was found (1966) in a pagoda that had been constructed in AD 751 at Bulguksa Temple, Gyeongju, Korea. While there has been some contention as to the origin of the text—whether it had originally been printed in China or in Korea—scientific analysis indicates that it was printed on paper typical of Korean product. This is now the earliest known surviving specimen of xylographic text printing. The frontispiece of the *Vajracchedikā*, however, remains the earliest extant woodcut illustration.

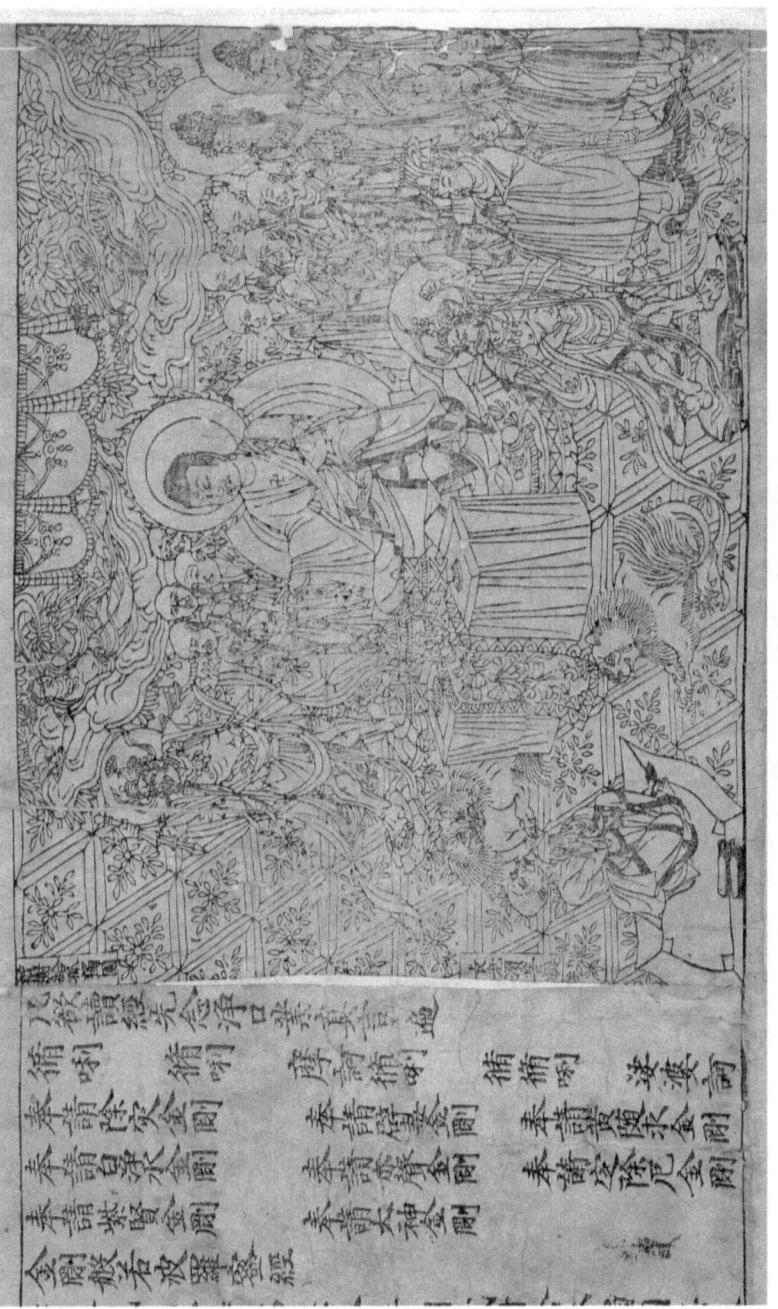

Fig. 1. Frontispiece of the *Vajracchedikā*, now held in the British Museum, London.

The first novel and significant departure in the age of the T'ang is the large number of ink factories springing up under the guidance of highly trained specialists. No less than twenty-five names of manufacturers of repute have been handed down from this epoch. In ancient times everyone was his own maker of ink, or the ink-maker was a man of no consequence. Under the T'ang, the business was taken out of private hands, and began to be industrialized and commercialized. With the vast expansion of the empire, governmental affairs increased in volume, and state correspondence assumed unparalleled proportions: Thus the government was compelled to maintain its own ink establishment, and we hear of an "ink official" (*mo kwan*, or *mo wu kwan*, "official of ink affairs") who was placed in charge of the government works, under instruction to send an annual supply of ink to the metropolis for the feeding of the administrative machine.

The most famous of these ink directors was Tsu Min, whose reputation was widely known throughout the empire, and whose best ink was prepared with a glue concocted from deer's antlers; his fame was so lasting that even in the fourteenth century his name was still forged on ink-cakes. Another ink expert, Wang Kiün-te, worked exclusively for the imperial ateliers, and very little of his products reached the general public, so that any of his inks in private possession were looked upon as veritable family treasures and heirlooms. He availed himself of two sets of chemical accessories, one consisting of pomegranate peels preserved in vinegar, buffalo horn, and sulphate of copper; the other being composed of the bark of *Fraxinus pubinervus*, pods of *Gleditschia chinensis*, sulphate of copper, and *Verbena officinalis* (Chinese *ma-pien-ts'ao*, "horse-whip herb").

It is worthy of note that another species of *Verbena* is used in India for the manufacture of ink. The T'ang emperors also maintained an ink factory at Jaochou is Kiangsi Province, where the slopes of the Lu Mountains (Lu Shan) were covered with pines. It is on record that the emperor Hüan Tsung (AD 713-755) sent every season 336 pellets of ink to two colleges which he had founded, the T'u shu fu and Tsi hien yüan; and the same monarch is said to have himself manufactured an ink with the juice of lotus-flowers (*Nelumbium speciosum*) blended with an aromatic powder, his product being known as "imperial ink" (*yü mo*). This is a very interesting fact in that it demonstrates that the occupation of ink-making was then a perfectly honorable and even dignified and exalted profession, and this is no wonder in a society where learning was so highly esteemed and worshipped.

Most of the ink manufacturers of the T'ang were men of culture, literati and officials, and thanks to their social status, their names have come down to posterity. This is in striking contrast to the fact that, as known to everyone, China has produced a long line of ingenious and clever potters and bronze founders, men of humble standing, and that hardly any of their names have survived. We know and admire their works, but are ignorant of their names; of the ink artisans we have their names, not their works. In China, so far as is known, no ink-cakes of the T'ang and Sung periods are preserved; among the oldest are those which have come down from the Ming Period. As regards the esteem in which the ink maker is held even in recent times, Du Halde states in his *Description de la Chine* (1738) that "everything which relates to writing is so reputable among the Chinese that even the workmen employed in making the ink are not looked upon as following a servile and mechanical employment."

In conformity with the specialization of work, the T'ang ink makers signed their pieces, covered them with inscriptions, and even provided them with a date. Thus under Kao

Tsung (AD 650-680) was issued an ink bearing the inscription, "Ink guarding the treasury [*chen k'u mo*], made in the second year of the period Yung-hui" (AD 651). One of these pieces is said to have weighed two catties. There was another ink inscribed, "Made by Li Ts'ao, second secretary of the Boad of Water Communication of the T'ang." Li Ts'ao was the ancestor of a family the members of which devoted their lives to the production of ink. They originated from Yi-shwi, but emigrated to Shōchou which forms the prefectural city of Huichou in Anhui Province; extensive pine-tree forests in that locality induced them to choose it as their domicile. This region has remained the principal seat of the ink industry until the present time.

Of the various Li, it was Li T'ing-kwei who attained the greatest fame. He was the "ink official" of the Southern T'ang dynasty (Nan T'ang, 937-975, also known as Kingdom of Kiang-nan, with Nanking as capital), and his products were regarded as the best in the empire and the goal which all subsequent manufacturers endeavored to reach; many also borrowed his name and counterfeited his ink-cakes. His originals usually bore an inscription consisting of four characters: *Li T'ing-kwei mo* ("ink of Li T'ing-kwei"); some also were dated, for instance, in the first year of the period Pao-ta (AD 943) offered by Shōchou and made by the official in charge of ink affairs, Li T'ing-kwei." His ink-cakes were four inches long, one inch wide, and one inch thick, and were frequently adorned with gilded dragons. They were as hard as metal and stone, and so hard and sharp that they could serve for smoothing printing blocks. Even after a hundred years, when rubbed, they still emitted an odor of camphor; and they yielded no sound in being rubbed on the stone, which the ancients always regarded as one of the criteria of good ink. A crackling sound was naturally caused by grittiness adhering in the ink, and good ink had to be free of any sandy matter. Other qualities demanded were that it should be deep black, light in weight, and extremely hard, its hardness being compared with that of jade.

In the palace the ink of Li T'ing-kwei was burnt, and the soot thus obtained was used as a paint for the eyebrows; it was hence styled "eyebrow-paint ink" (*hua mei mo*). His recipe was kept a secret and died with him. Ch'ao Shwo-chi, in his *Mo king*, states that he availed himself of twelve chemical ingredients, but is only able to name four of them; these are gamboge, the inspissated sap derived from incision into the bark of *Garcinia morella* or *G. hanburyi* (this beautiful reddish-yellow pigment is still used by Chinese draughtsmen and painters), rhinoceros-horn, genuine pearls, and the seeds of *Croton tiglium*. There is no doubt, of course, that lampblack of pine, as with all manufacturers of the T'ang period, formed the essential of his ink. He is also credited with the production of an ink of blue color.

Next in reputation to the ink of Li T'ing-kwei was that of Chang Yü from Yi-shwi, whose products were known as "tribute ink of Yi-shwi" (*Yi-shwi kung mo*). He lived toward the end of the T'ang period, and his inks were made in the shape of copper coins, which was rather inconvenient for rubbing them. Another noted ink artisan was Chu Fung, who worked for Han Hi-tsai, a scholar and minister of state. His atelier at Shöchou was known as the "hall where pine-trees are transformed" (*hoa sung t'ang*), and his ink was inscribed *Yüan chung-tse* or *Shö hiang yüe hia* ("musk moon box").

In the beginning of the T'ang dynasty (from AD 618) Korea (*Kao-li*) sent to the court of China an annual tribute of ink made from the lampblack of very old pine (*sung yen mo*) mixed with glue obtained from the antlers of the tailed deer (*mi lu*, *Cervus davidianus*). This Korean ink is described as black as if it were coated with varnish, glossy, and floating in water (*Wei lio*, Ch. 12, p. 1). It is sometimes asserted that the Koreans were the first who manufactured ink from pine lampblack, and that it was the Chinese who subsequently adopted the process; this conception of the matter, however, is not correct. The pine lampblack was utilized in China prior to the T'ang, probably as early as the third and fourth centuries AD, and the Koreans learned the whole technique from the Chinese; but the Koreans succeeded in perfecting the product to a degree unknown in China by employing a particularly suitable pine-wood well dried in the course of years and a specially fine hart's-horn glue. The latter was known to the Chinese in early times, and is looked upon as the finest kind of glue (those next in rank being derived from the hides of horse, cow, rodents, and rhinoceros); it is called white glue (*pai kiao*) or yellow bright glue (*huang ming kiao*), the latter designation being also applied to the ink thus prepared. It was naturally expensive, as this glue was hard to obtain. The Korean product elicited the admiration of the Chinese, and for some time its method of manufacture remained a secret to them. They made several endeavors to imitate the art of the Koreans, but only attained the desired result toward the close of the T'ang dynasty (about AD 900); yet under the Ming the secret of making this ink was lost (*T'ung ya*, by Fang I-chi, Ch 32, p.17).

Under the Sung (AD 960-1279) we hear of several manufacturers who availed themselves of antlers' glue; thus, for instance, Chang Kü-tsing. Others, like P'an Ku, who lived during the latter part of the eleventh century, and

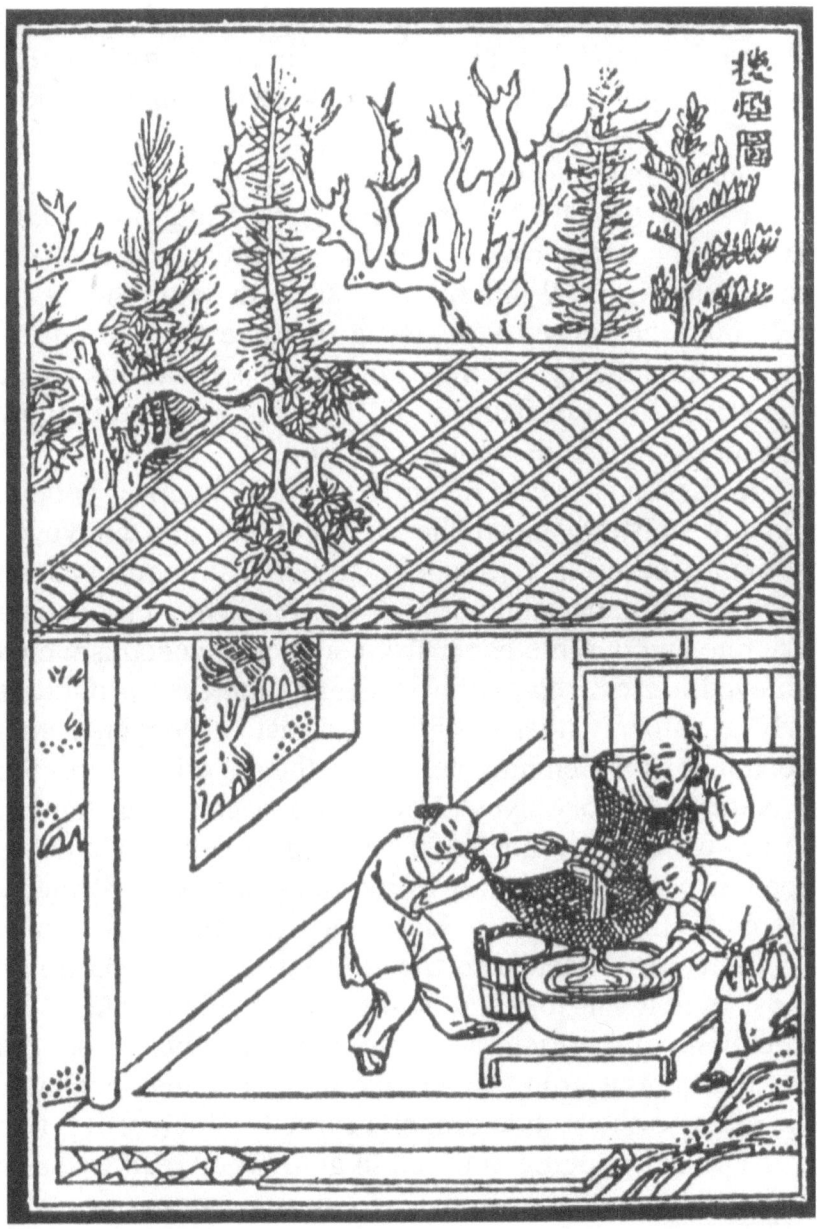

Fig. 2. Chinese ink-makers straining oil; from an old xylographic print.

who belongs to the most renowned ink manufacturers of the period, remodeled specimens of Korean ink by breaking them up, pounding the mass, and mixing it again with glue. P'an Ku is also noted for having used in his ink a sort of ivory-black made from bones.

The poet Su Shi, better know as Su Tung-p'o (1036-1101), prepared his own ink by using Korean charcoal and glue of the Kitan. When he lived as an exile on the island of Hainan, he caused P'an Heng to make for him an ink bearing the legend, "Hainan pine-tree charcoal, ink made after the method of Tung-p'o." This ink was presumably made according to the Korean method.

The Koreans were inventive and ingenious people, who not only advanced the cause of ink and conveyed its manufacture to the Japanese, but also improved on processes of paper-making and printing. They produced a very fine and durable paper from silkworm cocoons which achieved a great reputation in China, and they printed as early as 1403[3] with movable type cast of copper, actual specimens of which may be seen in the American Museum of Natural History, New York. One peculiar custom of Korea, which may not have been practiced in China or Japan, is particularly noteworthy: Ink-cakes were formerly offered by the emperor of Korea as a sacrifice to the gods, and there was attached to the Court a special department whose duty it was to manufacture this sacrificial ink. In the collections of the Field Museum, Chicago, there is an old ink-cake coming from Korea, of oblong, rectangular shape. The obverse shows two dragons carved in high relief, two large characters being outlined in gold. They read *kwo pao*

---

[3] Current scholarship dates the employment of movable metal type in Korea to as early as the 13th century. See further fn 18, p. 97.

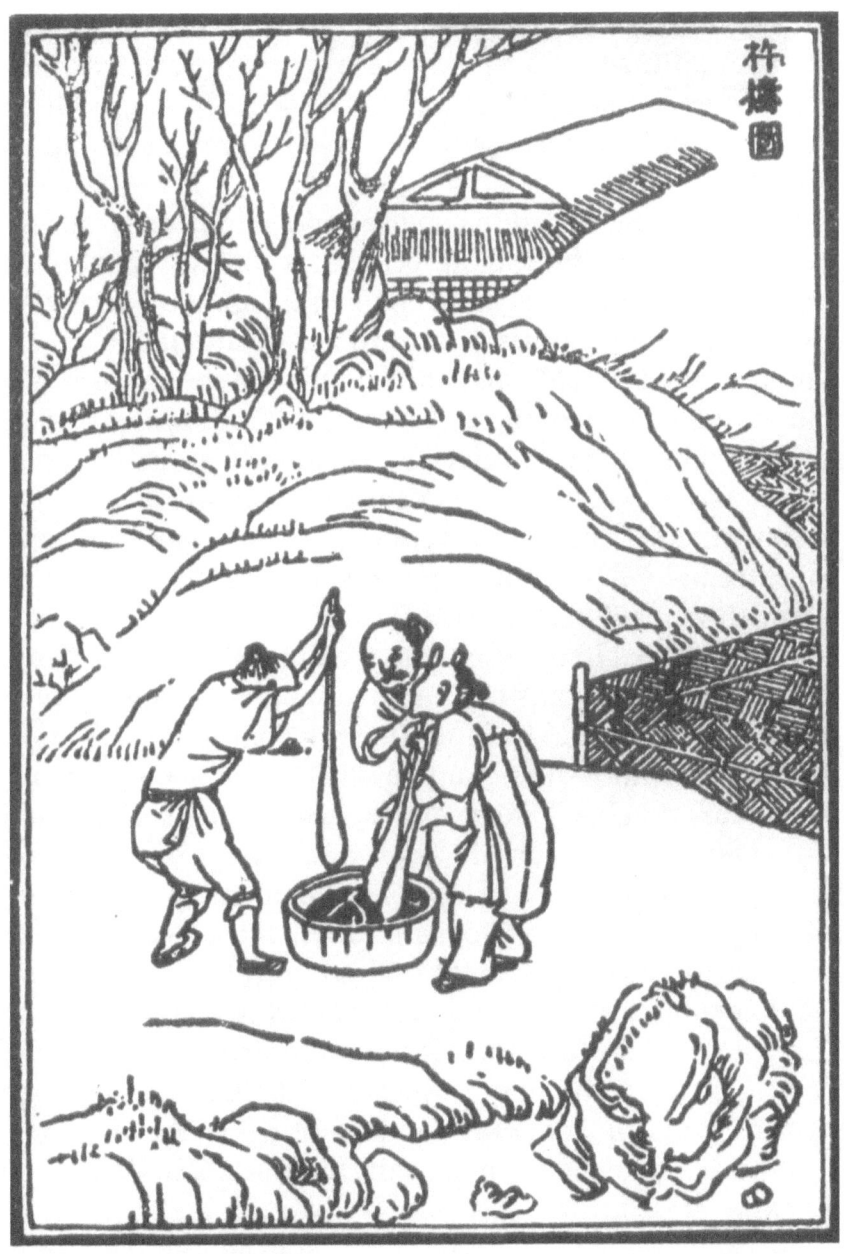

Fig. 3. Compounding Chinese ink, from an old xylographic print.

("treasure of the country, national treasure"). The reverse contains an inscription in Chinese to the effect that this ink was made in the Yung-lo period (1403-25) of the Ming dynasty; but whether manufactured in Korea or China is not stated.

Under the Sung a good many innovations as to detail were introduced, but the old principles virtually remained the same. Some improved on the lampblack, others on the glue, fish-glue was then first utilized, while others again devoted much thought and pains to the proper proportions and methods of bonding of the two. Shön Kwei, a native of Kia-ho in Hunan, produced a lampblack from charcoal of old pines which he blended with pine-resin and the sediments of varnish; this compound when burnt yielded an extremely fine lampblack which received the name "varnish-smoke" (*ts'i yen*).

The principal innovation that took place under the Sung was the substitution of vegetable, mineral, and animal oils for pine lampblack. The latter method continued in Anhui; the new method sprang up in the central and western provinces, notably Hunan and Szechuan. It is clear that pine lampblack, after all, was a comparatively costly matter, that pine trees were not available everywhere, and that the ever-increasing demand for ink must have resulted in a despoliation of forests and contributed its share to that lamentable state of deforestation which has proved so grave a calamity to China. It is not surprising, therefore, that in view of the huge expansion of literature and art, writing, printing, drawing, and painting in the glorious age the Sung, cheaper materials for the manufacture of ink were sought for.

Hu King-shun, a native of Ch'angsha (at that time called T'anchou) in Hunan, obtained lampblack by burning the oil of *Aleurites vernicia* Hassk. (=*A. cordata* Steud., formerly *Elaeococca verrucosa* and *Dryandra cordata*); this was called "*t'ung* flower smoke" (*t'ung hua yen*) and proved a great success. His ink was much prized by painters for drawing the pupils of the eye. This oil, commercially known as *t'ung* oil or wood oil, is still accorded preference to any other in the modern manufacture of ink; a hundred catties of it yield eight catties of pure lampblack. Next in appreciation come the seeds of *Sterculia platanifolia* (*wu-t'ung*), which contain a good oil used in making ink. It is a stately, ornamental tree, frequently planted in the courtyards of temples and houses, its large leaves affording an excellent shade. It may be mentioned here that the juice from the crushed seeds is rubbed into gray hair, with the reputed virtue of causing the gray to fall out and the new hair to come in black.

In the period between AD 1067 and 1084, Chang Yü manufactured what became known as *yü mo* ("imperial ink"), as he presented his product to the Court. He used lampblack made from oil blended with musk, camphor, and gold-leaf, and is said to have been the first who availed himself of oil. His product was called *lung hiang tsi* ("dragon fragrance compound").

The oil-combustion ink seems to have at first roused the suspicion of some people, for an anecdote has it that when P'u Ta-shao produced his oil ink, people anxiously asked him how such ink could be strong and lasting; he assuaged the skeptics by responding that half of it contained pine lampblack, without which it could not be permanent. Whether this was correct or merely a subterfuge we do not know; it may be that such a "half-and-half" composition was really made, but it is certainly untrue that pine lampblack is necessary to give ink permanency. P'u Ta-shao was a native of Lang-chung, which forms the prefectural city of Paoning in Szechuan Province, and his ink was widely used by scholars and offered as present to the throne.

K'ou Tsung-shi, author of the herbal *Pen ts'ao yen i*, written in AD 1116 under the Sung, has the following interesting account:

> "Ink is made from the black produced by the smoke of pine-trees. Our contemporaries manufacture a sham product from the ashes of grain stubbles, which should not be used. Only pine-soot ink is serviceable in the *materia medica*. Solely distant smoke is fine and yields an excellent product; the coarse one should be discarded. At present Korea dispatches ink to China with every mission of tribute, but the ingredients of this ink are not known, nor is it beneficial as a medicine. In Fuchou and Yen-ngan fu [both in Shensi Province] there is kerosene [*shi yu*, 'stone oil']; the smoke emanating from it is very thick; the black produced by it can be made into ink. It has a black gloss like varnish, but cannot be employed medicinally."

The use of this petroleum ink inaugurated under the Sung has already been pointed out. It is also mentioned by Shön Kwa, who wrote the *Mong k'i pi t'an* in the middle of the eleventh century. He states that natives of Yen-ngan in Shensi burnt petroleum in their lamps, swept the lampblack together, and made it into ink, which had a deep black brilliancy like varnish, and which was superior to ink made from pine-resin.

There seems to have been some good reason for K'ou's cautioning against the medicinal employment of Korean ink; for it is on record that flies when sucking the juice of it would die, but by dint of what poison remained unknown. The same author alludes to the use of oak-galls as a hair-dye and ink for domestic purposes; but this is an isolated instance, and such ink has never become popular.

Under the art-loving emperor, Hui Tsung (1100-25), attempts were made to mix lampblack with storax, the sweet-scented resin of *Liquidambar orientalis* and *L. altingiana* (cf. *Sino-Iranica*,[4] p. 456). It is said that an ounce of this ink was worth a pound of gold, and that efforts to imitate it failed.

Of oils of animal origin, preference was given to lard. On this point the Jesuit Louis Le Compte, at the end of the seventeenth century, writes:

"China ink is not so difficult to make as people imagine; although the Chinese use lampblack drawn from divers matters, yet the best is made of hog's grease burnt in a lamp: they mix a sort of oil with it to make it sweeter, and pleasant odors to suppress the ill smell of the grease and oil. After having reduced it to a consistence, they make of the paste little lozenges, which they cast in a mould; it is at first very heavy, but when it is very hard, it is not so weighty by half, and that which they give for a pound, weighs not above eight or ten ounces."

Fan Ch'eng-ta, in his interesting work *Ling wai tai ta*, which deals with the geography and products of southern China, and which was written in AD 1178, has the following note on ink (Ch. 6, p. 2b):

"In Jungchou [prefecture of Wuchou, Kwangsi Province] there is an abundance of large pine-trees from which the inhabitants manufacture ink. Good qualities are sold by the pound [catty] which is worth two hundred copper coins only. The merchants raise the money jointly for the purpose and sell

---

[4] This refers to Laufer's own work, *Sino-Iranica. Chinese Contributions to the History of civilization in Ancient Iran, with special reference to the history of cultivated plants and products*. Field Museum, Anthropological Series, vol. XV, 1919, No.3.

the stock as a whole. The ink of Kiao-chi [Tonking][5] although not very good, is not quite worthless either. The people barter their ink for horn, ink-slabs, and writing-brushes which they carry suspended from their loins."

John Francis Davis (*China*, Vol. II, 1857, p. 180) writes, "Chinese ink has been erroneously supposed to consist of the secretion of a species of sepia, or cuttle-fish. It is, however, all manufactured from lampblack and gluten," etc. Likewise J. Dyer Ball (*Things Chinese*, 4th ed., 1903, p.105) asserts, "A curious idea was prevalent at one time in the west that the so-called Indian ink was prepared from the coloring matter of the cuttle-fish, instead of being made from lampblack as its principal ingredient." These statements are only partly true. The fact remains that the Chinese formerly made use of sepia also, though to a limited extent.

The cuttle-fish, styled by the Chinese "*wu-tse* fish," was already known to T'ao Hung-king (AD 452-536), a celebrated physician and alchemist, who says that this creature carries ink in its belly, and that this substance was used as ink in his time. It is therefore called also "ink-fish" (*mo yü*). It is popularly believed to be a transformation of a crow, and another legend has it that it owes its existence to the emperor Ts'in shi when he dropped his writing outfit into the sea. This cephalopod is met with all along the coast of China and forms an article of trade at Ningpo and Wenchou in Chekiang. Large quantities are eaten dried or pickled, or taken as a tonic. The small bag of inky fluid situated near the liver of the cuttle-fish is understood to be its gall. The preparation of sepia as a pigment, however, was never understood; and the employment of the secretion as ink has never attained popularity, as it fades within a few years.

---

[5] Both of these are archaic names for present-day northern Vietnam.

Peter Mundy, who visited Canton and Macao in 1637, toward the end of the Ming dynasty, is one of the earliest European travellers who has recorded the use of ink in China. He noted that "they all write with pencills and blacke and red incke made into dry paste which they distemper with water when they will use it." He likewise gives a rough sketch showing a Chinese at his desk engaged in writing and an ink-well "which holds his incke, the one side containing blacke, the other redde; little partitiones with water where he dippes his pensill and so tempers his incke."

A formula for making ink from orpiment is given as follows:

"Ochre should be pounded into a very fine powder over which water is swiftly poured to clarify it. The water is poured out. Take the bark of *ts'in-p'i* [*Fraxinus pubinervus*], fruit of Gardinia, and pods of *Gleditschia chinensis*, one-tenth of an ounce of each, one grain of the seeds of *Croton tiglium* after removal of the skin, mix this with half an ounce of bright yellow Kwangtung glue made from ox-hides, boil this mass, mix it with the ochre and form it into cakes; or let dry in the shade and use it thus."

I add two recipes from Du Halde's *Description of China* (1738) in which he says:

"are taken from the Chinese and which perhaps may suffice to make the ink of a good black, which is looked upon as an essential property. Burn, they say, lampblack in a crucible, and hold it over the fire until it has done smoking. In the same manner burn some horse-chestnuts, till there does not arise the least vapor of smoke. Dissolve some gum tragacanth; and when the water in which the gum is dissolved becomes of a proper consistence, add to it the lampblack and horse-chestnuts, and stir all together with a spatula. Then put this paste into moulds; and take care not to put in too much of the horse-chestnut, which would give it a violet-black.

"A third receipt, much more simple, and easier to be put in practice, has been communicated to me by P. Contancin, who had it from a Chinese, as skilful in this matter as anyone can be expected to be; for we ought not to suppose that ingenious workmen discover their secret; on the contrary, they take the greatest care to conceal it, and make a mystery of it, even to those of their own nation.

"They put five or six lighted wicks into a vessel full of oil, and lay upon this vessel an iron cover, made in the shape of a funnel, which must be set at a certain distance, so as to receive all the smoke. When it has received enough, they take it off, and with a goose feather gently brush the bottom, letting the soot fall upon a dry sheet of strong paper. It is this that makes their fine and shining ink. The best oil also gives a luster to the black, and by consequence makes the ink more esteemed and dearer. The lampblack which is not fetched off with the feather, and which sticks very fast to the cover, is coarser, and they use it to make an ordinary sort of ink, after they have scraped it off into a dish.

"When they have, in this manner, taken off the lampblack, they beat it in a mortar, mixing with it musk, or some odoriferous water, with a thin size to unite the particles. The Chinese commonly make use of a size which they call *niu kiao* ('size of neat's leather'). When this lampblack is come to the consistence of a sort of paste, they put it into moulds, which are made in the shape they design the sticks of ink to be. They stamp upon the ink, with a seal made for that purpose, the characters or figures they desire, in blue, red, or gold color, drying them in the sun, or in the wind."

An Arabic author, Abu'l Faraj (AD 988), writes that the Chinese have an ink composed of a mixture resembling Chinese grease. He pretends to have seen specimens in the form of tablets representing the image of the emperor, and adds that a piece like this will last for a long time. This notice is of interest in that it shows that ink-cakes were embellished with designs at that early date; yet, the Arabic writer must have been mistaken as to the subject of the picture, for the emperor's portrait was never allowed to be turned to so profane a purpose, nor has it ever been customary in China to circulate an emperor's portrait during his lifetime. No imperial portrait appears on any Chinese coin, the only exception being in recent times that of the emperor Kuang-su on the Tibetan rupee (coined in Ch'engtu for the purpose of counteracting the influence in Tibet of the Anglo-Indian rupee bearing the portrait of Queen Victoria).

In Chinese records we read of representations of dragons on inks of the T'ang and Sung periods, but the Ming dynasty (1368-1643) was the great era when ink manufacturers appealed to noted artists for designs and pictorial representations to be applied to their products. Many specimens of this art have survived, and are eagerly bought by Chinese collectors. Some ink manufactures published books containing illustrations of all pictures placed on their inks. Two of these works are especially prominent. One is the *Fang shi mo p'u*, published in 1588 in six volumes by Fang Yü-lu, who manufactured ink at Shōhien, forming the prefectural city of Huichou in Anhui Province. Most of his illustrations were contributed by an eminent artist, Ting Yün-p'eng, who usually signs his pictures Nan-yü. He was a native of Hiuning in Anhui, and thus in close contact with his countryman, Fang Yü-lu. His designs are highly artistic, exceedingly fine, and well drawn, and represent a microcosm of Chinese mythology, as well as a thesaurus of art-motifs. Nan-yü revived several ink designs of which merely a literary reminiscence was preserved; thus an ancient text says that there formerly was an ink of nine children (*kiu tse mo*), which implied the wish that the owner of the ink may be blessed with an abundance of progeny.

Hence Nan-yü introduced the drawing of nine boys engaged in play, flying a kite, riding a hobby, manipulating a movable puppet, beating gongs and a drum. A goodly number of ink pieces are fashioned in the shape of archaic jade ornaments; others, in the shape of a peach-leaf, conch-shell, or bell; others are adorned with designs of palaces, terraces, landscapes, flower-pieces, birds and quadrupeds, the eight famous steeds of King Mu, stargods and other deities of ancient lore, or Buddhist emblems accompanied by Indian scripts. This, of course, is not the place to give a detailed account of the variety and significance of these designs; suffice it to call attention to this artistic development of ink-cakes which is a unique phenomenon in the history of art.

The other work is the *Ch'eng shi mo yüan* in twelve volumes, issued by Ch'eng Kün-fang (or Ta-yo, but commonly called Kün-fang) between 1594 and 1606, an excellent copy of which, printed on Korean paper, was secured by the writer for the American Museum of New York in 1901, another for the John Crerar Library of Chicago in 1908. This belongs to the finest and most artistic examples of Chinese book-making. It contains 385 cuts accompanied by explanations, essays, and poetry, the handwritings of the authors being reproduced in facsimile. One of the interesting features of Ch'eng's work is that it contains four Christian pictures contributed by the celebrated Jesuit Matteo Ricci, who arrived at Macao in 1582, and who died in 1610; the engravings are accompanied by an essay and explanations written in Chinese by Ricci himself and reproduced in facsimile. They have been published by the writer in an essay entitled "Christian Art in China."[6]

---

[6] "Christian Art in China," *Mitteilungen des Seminars für orientalische Sprachen*, vol. XIII. Berlin, 1910.

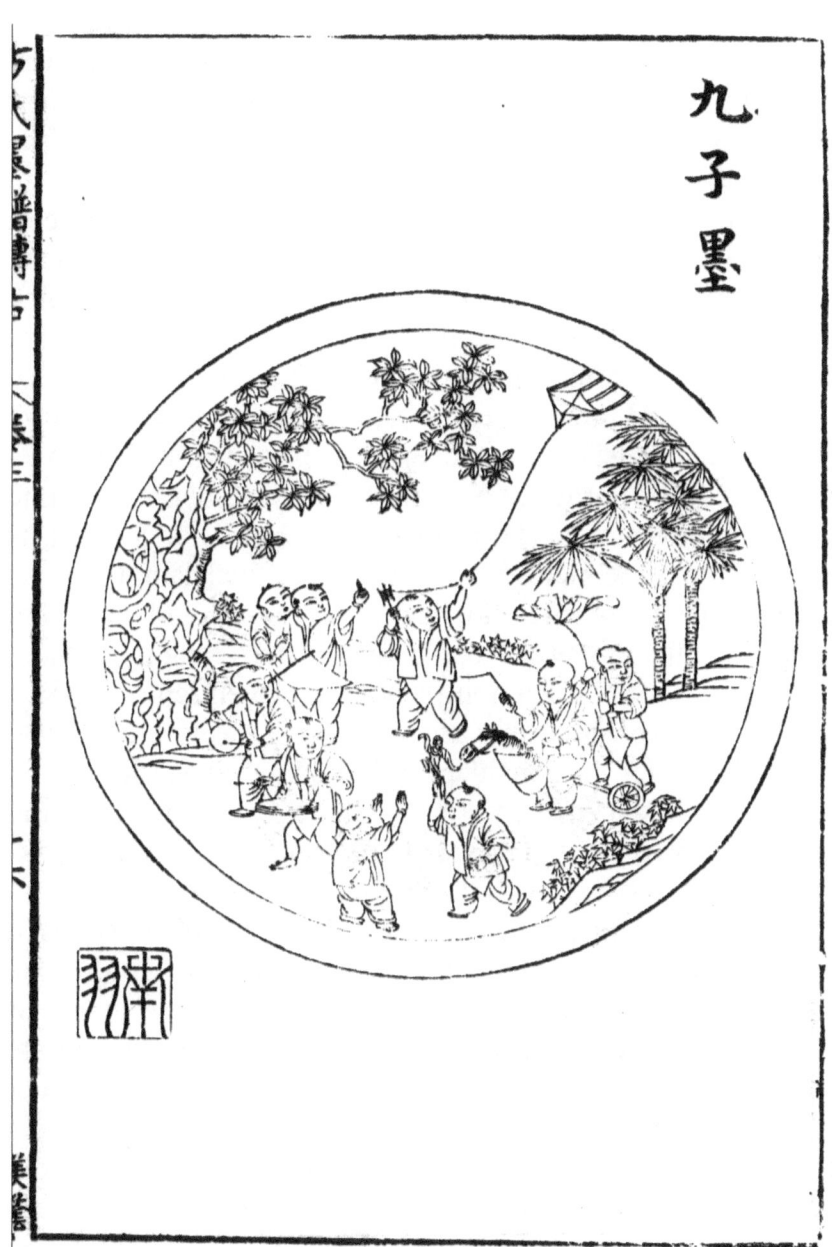

Fig. 4. Illustration of an ink-cake 'nine children' design by Nan-yü, from the *Fang shi mo p'u*.

The Newberry Library of Chicago of Chicago received from the writer a similar Japanese book, entitled *Ko-bai-en bokufu* [in Chinese: *Ku mei yüan mo p'u*, ('Collection of the inks of the Old Plum-tree Garden')], published in 1773, in five volumes. The first of these contains a number of prefaces, two volumes are filled with eulogies of ink facsimiled in the handwritings of the authors, and two others are occupied by engravings of ink designs. In the main these follow the models of their Chinese prototypes, but do not quite display their vigor and power; yet, in some cases, they also exhibit original subjects, for instance, ink-cakes in the form of a daimio's suit of armor, bow, and boots. One bears the name of the illustrious Li T'ing-kwei of the T'ang; another is adorned with the picture of the Envoy of the Black Pine, the spirit or genius of ink who, according to a legend, appeared one day to the emperor Hüan Tsung of the T'ang dynasty.

Ink is a favorite gift among scholars and gentlemen, and for this purpose special boxes are made up in a very elegant and tasteful manner. In consideration of some courtesies which had been shown his son, a high Chinese official of Peking once presented the writer with a box containing eighteen ink-cakes, each adorned with the portrait of a famous scholar in low relief, his name being written in gold, the reverse of each cake bearing a stanza accompanied by seals in gold. Nine cakes are arranged in a finely lacquered tray, the inside of which is mounted with yellow silk. Each tray has its separate cover likewise lacquered black, decorated with a border design in gold, and inscribed with four large characters in heavy gold (*fang ku tsang yen*, "in imitation of ancient lampblack"). The two lacquer boxes fit into a cardboard case mounted on decorated cloth in the style of bookbindings, and the whole presents the appearance of a book. Other such series are exquisitely adorned with celebrated landscapes or

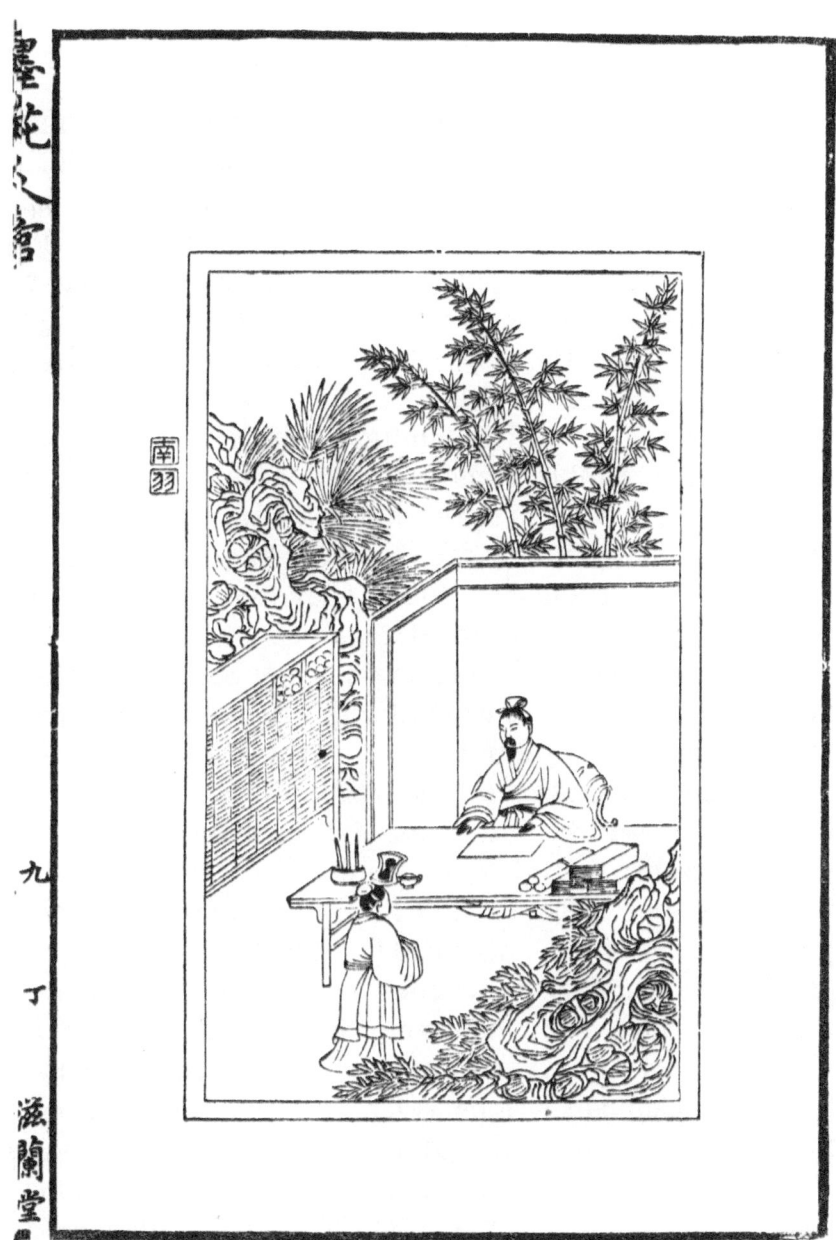

Fig. 5. Illustration of an ink-cake design depicting a scholar at his desk, from the *Ch'eng shi mo yüan*.

mountain scenery, or show all the stages in the process of tillage and weaving.

For preserving ink wrapping it up in a leopard-skin is recommended by Fung Chi in his *Yün sien tsa ki* (Ch. 1, p.7), written in the beginning of the tenth century. The remedy may be efficient, but leopard-skins were not within everyone's reach, nor hardly ever available in a quantity sufficient to go round. The average man therefore had to be content with a plain box in which mugwort (*Artemisia vulgaris*) was placed as a means of preserving the ink. The main point is that it should not be exposed to sunlight, which would cause it to crack and crumble to pieces. The box therefore must be tight-fitting. As early as the T'ang period special boxes were turned out for keeping ink, and great luxury was displayed in them under the Sung and Ming. They were made of a kind of sandalwood (*Pterocarpus santalinus*), ebony, or *nan-mu* (*Persea nanmu*), a valuable timber of Szechuan, which does not easily rot, and which for this reason is much used for buildings and furniture. The wood was inlaid with jade plaques derived from the court-girdles of the T'ang period or with designs of hydras, tigers, and genre-scenes carved in jade; it was also coated with red and black lacquer.

The prominent qualities of Chinese ink are well known. It produces, first of all, a deep and true black; and second, it is permanent, unchangeable in color, and almost indestructible. Chinese written documents may be soaked in water for several weeks without washing out. It is safely used to mark linen. In documents written as far back as the Han period in the beginning of our era, and discovered under the sand of Turkestan, the ink is as bright and well preserved as though it had been applied but yesterday. The same holds good of the productions of the printer's art. Books of the Sung, Yüan, and Ming dynasties have come down to us with paper and type in a perfect state of preservation. Above all, we owe to the perfect ink of the Chinese many masterpieces of the brush in black and white and that charming art of monochrome drawing. Li Kung-lin or, as he is better known, Li Lung-mien, was one of the greatest masters of the line who ever lived, and the inspirations of his genius were merely expressed by black ink on white paper; he made the ink live and speak, drawing his line in hundreds of shades.

Speaking of the qualities of China ink, the Jesuit Louis Le Compte (*Memoirs and Observations made in a Late Journey through the Empire of China*, p. 192, London, 1697) wrote at the end of the seventeenth century, "This ink is shining, extreme black, and although it sinks when the paper is so fine, yet does it never extend further than the pencil, so that the letters are exactly terminated, how gross soever the strokes be. It has moreover another quality, that makes it admirable good for designing, that is, it admits of all the diminutions one can give it; and there are many things that cannot be represented to the life without using this color."

Naturally Europe endeavored to rival the Chinese competition. Attempts were made in France at an early date to imitate China ink. Father Le Compte writes with reference to this subject, "It is most excellent, and they have hitherto vainly tried in France to imitate it; that of Nanking is most set by: And there are sticks made of it so very curious and of such sweet scent that one would be tempted to keep some of them though they should be of no use at all." Likewise Du Halde, in his *Description of China* (London, 1738) observes, "The Europeans have endeavored to counterfeit this ink, but without success. Painters and those who delight in drawing know how useful it is for tracing their sketches, because they can give it what degree of shade they please." And still in 1869 P. Champion (*Industries anciennes et modernes de l'empire chinois*, p. 129) wrote that many attempts were made in France to manufacture Chinese ink, but that the results have never been entirely satisfactory, and that the ink of Chinese origin, superior in quality to the French products, has always been preferred by the draughtsmen. Chinese ink can be made only in China, and will never be equalled anywhere else.

Huichou in Anhui Province, where ink manufacture was so successfully initiated under the T'ang, is the high seat of the industry also at the present day. It is still of interest to read Du Halde's account of the Huichou factories. He writes:

"We are assured that in the city of Huichou, where the ink is made which is most esteemed, the merchants have great numbers of little rooms, where they keep lighted lamps all day; and that every room is distinguished by the oil which is burnt in it, and consequently by the ink which is made therein. Nevertheless many of the Chinese believed that the lampblack, which is gathered from the lamps in which they burn oil of gergelin [sesame], is only used in making a particular sort of ink, which bears a great price, but considering the surprising quantities vended at a cheap rate, they must use combustible materials that are more common, and cheaper.

"They say that lampblack is extracted immediately from old pines, and that in the district of Huichou where the best ink is made, they have furnaces of a particular structure to burn these pines, and to convey the smoke through long funnels into little cells shut up close, the insides of which are hung with paper. The smoke being conveyed into these cells sticks to every part of the wall and ceiling, and there condenses itself. After a certain time they open the door, and take off a great quantity of lampblack. At the same time that the smoke of these pines spreads itself in the cells, the resin which comes out of them runs through other pipes, which are laid even with the floor.

"It is certain that the good ink, for which there is a great demand at Nanking, comes from the district of Huichou, and that none, made elsewhere, is to be compared with it. Perhaps the inhabitants of this district are masters of a secret, which it is hard to get out of them. Perhaps also the soil and mountains of Huichou furnish materials more proper for making good lampblack than any other place. There is a great number of pine-trees; and in some parts of China, these trees afford a resin much more pure, and in greater plenty, than our pines in Europe. At Peking may be seen some pieces of pinewood which came from Tartary, and which have been used for above these sixty years. Nevertheless, in hot weather, they shed a great quantity of big drops of resin resembling yellow amber. The nature of the wood which is burnt contributes very much to the goodness of the ink. The lampblack which is got from the furnace of glasshouses, and which the painters use, may perhaps be the properest for imitating Chinese ink.

"As the smell of the lampblack would by very disagreeable, if they were to save the expense of musk, which they most commonly mix with it; so by burning such drugs, they perfume the little cells, and the odors mixing with the soot, which hangs on the walls like moss, and in little flakes, the ink they make thereof has no ill scent."

Huining, in the prefecture of Huichou, is now the centre of the ink industry of Anhui Province. From there the whole of China is supplied with ink. That of the family Hu K'ai-wen injoys a special reputation.[7] It sells its product from 300 copper coins up to 48 ounces of silver (taels)[8] for the pound which includes 30-32 cakes of medium size. The ink of the first class is made by varnish and sesame oil; that of the second class, by sesame oil and lard; that of the third class, by the oil of colza; and that of the fourth class by t'ung

---

[7] For a fine example of the product of the Hu k'ai-wen family studio, see back cover illustration.

[8] About $650 US Dollars in 2019 currency.

oil. Each of these substances makes inks of very different quality, according to the number of lamps and the degree of slowness of combustion. Two good workmen can turn out eighty pieces daily, each half a pound in weight (cf. H. Havret, *La Province du Ngan-hoei*, p.38).

In 1863, according to S. Wells Williams' *Chinese Commercial Guide*, the finest ink was priced as high as Mexican $5 a catty[9] (about 1⅓ lbs.), common sorts ranging from $0.40 to $1.50. The boxes destined for export to Europe usually contained a hundred cakes.

The usual method of printing books in China and Tibet is that by means of wooden blocks. For this purpose, the manuscript is first written on thin paper by a professional calligraphist. This paper is pasted over the finely planed block with the characters turned face downward, the thinness of the paper displaying the writing perfectly through the back. Then commences the engraver's work, who chisels down the surface of the block around the characters, so that the writing in negative stands out in relief. In this state, the blocks go to the printer, who lightly rubs ink over them with a round brush of coir-palm fibre, places a sheet of paper on them, and takes the impression by passing another brush over.

In regard to printer's ink, old Du Halde (1738) has supplied the following information from the reports of the Jesuits:

> "The ink which they use of for printing is a liquid, and therefore much more convenient than that which is sold in sticks. To make it, you must take lampblack, pound it well, expose it to the sun, and then sift it through a sieve. The finer it is, the better. It must be tempered with Aqua-vitae till it comes to the consistence of size, or of a thick paste, care being taken that the lampblack may not clot. After this it must be mixed with a proper quantity of water, so that it may be neither too thick, nor

---

[9] About $40 US Dollars in 2019 currency.

too thin. Lastly, to hinder it from sticking to the fingers, they add a little neat's leather glue, probably of that sort which the joiners use. This they dissolve over the fire, and then pour on every ten ounces of ink almost an ounce of glue, which they mix well with the lampblack and Aqua-vitae, before the water is added to them."

De Guignes (*Voyages à Peking 1784-1801*, Vol. II, p. 229) gives the following note on printer's ink:

"For purposes of printing they avail themselves of a particular and rather fluid ink. It is made from lampblack finely ground, which is passed through a very fine sieve. It is then soaked in rice wine, and when it has the consistency of a pap, glue is added at the dose of an ounce for ten ounces of lampblack. The whole is mixed together, with the addition of the necessary quantity of water."

For printing-ink, S. Wells Williams informs us, the lampblack is mixed with strained congee or a vegetable oil; and when the paste is properly dried, it is kneaded on a slab and cut into strips shaped like wrought nails. The printers grind it, or dilute it in oil as they use it, laying it on the wooden blocks with a brush made from the bark of the coir-palm.

At the present time [ca. 1930] large quantities of printer's ink are imported into China from Japan, and that of Japan is made according to European methods.

A specimen of powdered printer's ink obtained by the writer in 1910 from a Tibetan monastery, but presumably of Chinese origin, was examined by Henry W. Nichols, associate curator of geology in the Field Museum, Chicago. It was actually used with good success in the Museum's printing office. Mr. Nichols reports as follows:

"I have made a qualitative chemical examination of the Tibetan ink powder submitted. The material is a nearly black powder of medium coarseness. A sizing test shows 30% retained on a twenty mesh sieve and only 10% fine enough to pass a hundred mesh sieve. The powder is evidently to be reground before used for ink. Under the microscope some of the larger particles take the form of broken fragments, others show spherical and botryoidal surfaces covered with small projecting points. Still others have a rough cellular appearance like that of clinkers and cinders from a coal fire. There is no trace of organic structure apparent.

"The ink is composed principally of carbon. When it is burned, the ash is too great in quantity for charcoal and too little for bone or ivory black. The ash is principally phosphate of calcium with smaller quantities of iron, alumina, silica, and undetermined elements. The ink cannot be a soot like lampblack or carbon black on account of the coarseness. The quantity and character of the ash show that it cannot be a pure charcoal nor a pure bone black.

"The examination indicates that this ink is similar to the older form of drop black except that the process of manufacture has been carried one step further than is the case with bone black. Charred ivory, teeth or the denser parts of bone has been mixed with charcoal and finely ground. The powder has been cemented into a cake with glue, gum or some similar substance. The cake thus formed has been ground to a coarse powder which has been recharred. Heath and Milligan described the old style of drop black thus: Drop Black, as the name implies, was first placed on the market in the form of small lumps or drops and consisted of various mixtures of animal and vegetable blacks ground to a fine powder with water, mixed with a little gum and then moulded into drops and dried. The Chinese material differs from this in that the moulded drops or cakes have been powdered and reburned. The object of adding bone black to the charcoal is to improve the color."

As any substance found in nature and any artifact, ink also has invaded the *materia medica* of the Chinese. In the *Pen ts'ao kang mu*, written at the end of the sixteenth century and regarded as the standard herbal of the late Ming and Manchu periods, ink is described as astringent, diuretic, emmenagogue and vulnerary in its qualities. It is recommended as an application to the eye when irritated by the presence of foreign bodies. Not so long ago, and perhaps still at present, stale ink was administered as a kind of paint for daubing over tumors and swellings of all kinds, also for treating ulcers and wounds. This does not appear so bizarre if we remember that similar practices prevailed among us. Thus Francesco Carletti (*Ragionamenti sopra le cose da lui vedute ne' suoi viaggi*, Vol. I, p. 84), who visited Peru in 1595, relates that the wound caused by an injurious insect called *higna* when removed from the skin was healed by the application of a little ink.

Du Halde writes,

"When the ink has been preserved a long time, it is then never used for writing, but becomes, according to the Chinese, an excellent and refreshing remedy, good in the bloody flux, and in the convulsions of children. They pretend, that by its alkali, which naturally absorbs acid humors, it sweetens the acrimony of the blood. The dose, for grown persons, is two drachms, in a draught of water or wine."

At the end of the eighteenth century de Guignes even recommended old Chinese ink for hemorrhages and the stomach, provided that it is of superior quality. This effect, he comments, is not surprising, as it is combined with *ngo-kiao* or glue from asses' skins, which is a supreme remedy in blood-vomiting. According to de Guignes' description, asses' glue would enter the composition of every ink, which, of course, is not true.

Writing-brush and ink have become so essential requisites and attributes of scholarship that *pi mo* ("brush and ink") has developed into a term denoting literature. The phrase "he talks of nothing but pen and ink" means that his hobby is literature. "Eating ink" is a common phrase for studying, from the habit of the Chinese of putting the writing-brush into the mouth in order to give it a fine point. The question addressed to a scholar, "how many years' ink have you eaten?" Means as much as "how long have you been studying?" A skilled writer is said "to scatter ink and make pearls."

Under the first emperor of the Liang dynasty (AD 502-556) candidates who failed in the examinations for the degree of *siu-ts'ai* were made to drink long draughts of liquid ink.

Ink has naturally entered into proverbial sayings also. "It is not the man who rubs (wears out) the ink, it is the ink which wears out the man,"—by his application to study. "He who is near ink gets black; he who goes near vermilion will make himself red." Written notes are regarded as preferable to memorizing. This is expressed by a proverb which says, "The palest ink is better than a capacious memory"; also quoted in the form, "A clever memory is not equal to a clumsy brush."

Poetical names for ink are "black metal," "dark incense," "black-jade ring."

The Chinese never keep liquid ink in bottles or ink-wells, but prepare only as much as they actually need at a time. For this purpose they avail themselves of a slab of marble or other stone which has a small rounded cavity at one end. A few drops of water are poured over the finely polished surface, and the stick or cake of ink is gently rubbed against it, the ink flowing into the cavity Many ink-sticks are provided with a rounded notch at the lower end to secure a firmer hold for the finger, while the upper part to be rubbed is rounded; in this manner one avoids confounding the two ends, as the wetted portion will naturally leave black spots on the fingers. The marble, before being used, must be carefully washed, so that no trace of old ink remains upon it; for even a small particle of old ink adhering to it is said to spoil both the marble and the fresh ink. The marble should not be cleaned with hot water or cold water just drawn from a well, but with water that has been boiled, and has grown cool again. The selection of the proper materials for ink-pallets and their preparation and carving has developed into a science in itself, and this subject has called forth a literature as exuberant as that on ink.

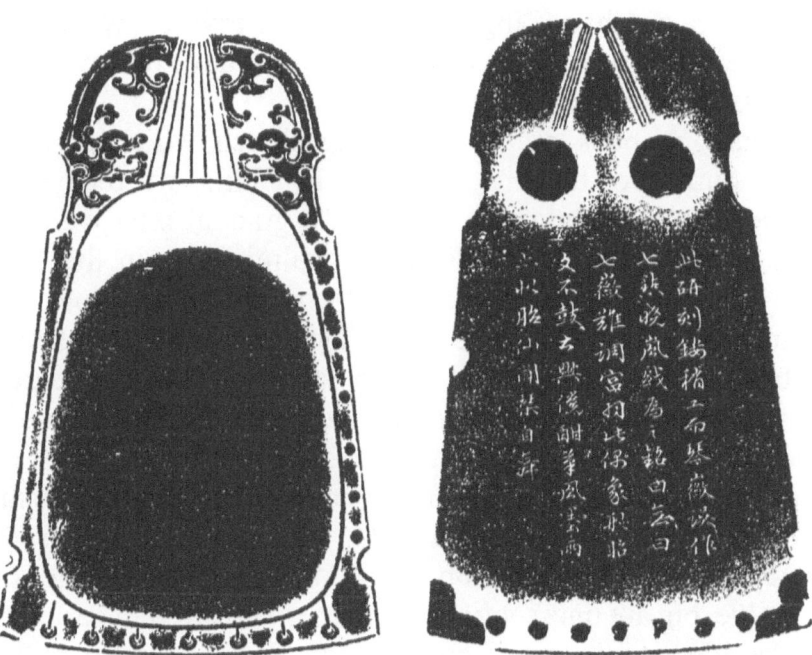

Fig. 6. Rubbing of an ink-stone in the form of a lute, Song dynasty, after R. H. van Gulik, *Mi Fu on Ink-stones*, Bangkok 2006.

White jade makes the finest ink-stones, and there is a specimen in the Field Museum in the form of a well frame, where the cavity is suggestive of a well—a veritable ink-well. Another very ancient specimen of cast iron is provided with a lower compartment for heating water with charcoal. A peculiar fad came into vogue during the eighteenth century to convert the ancient roofing tiles from the palaces of the Han emperors into ink-pallets. The field Museum has a very fine stone ink-slab of the T'ang period with carved figures of a lion and lioness of realistic style. The Twan-k'i stone has been famed from remote times for its use as ink-pallets; Twan-k'i is an old name for Te-k'ing, a district in the prefecture of Chao-k'ing in Kwangtung Province, where the quarries are situated. Several monographs have been written on this stone alone.

Glancing back at the preceding sketch which gives a mere outline of the history of ink in China, we recognize a constant ascending development, a gradual improvement and perfection of methods finally culminating in the best and most durable ink produced in the world. In principle, the composition of all Chinese inks is identical: the fundamental substance making the ink is lampblack from whatever source it may be derived; this is compounded with glutinous matter, the glue serving the purpose of uniting the fine particles of carbon and fixing the ink on paper by means of the brush. The perfumes sometimes added, like musk, camphor or patchouli, have the function of hiding the unpleasant odor of the glue, but are unessential. The numerous different varieties and grades of ink depend upon the fineness and quality of the lampblack, the quality of the glue, the proper proportions of the two, the process adopted in mixing them, and general methods of manufacture usually kept very secret. Another differentiation comes in from the addition of accessory vegetable, mineral, or chemical substances, which seem to vary in the hands of every manufacture. For these reasons it is obvious also that an absolutely correct description of the process which would hold good or be typical of all factories cannot be given. Few foreigners had occasion to obtain access to them, and still fewer possessed the technical knowledge to describe exactly what was going on. A French chemist, Paul Champion, has studied the process at Shanghai and Hankow toward the middle of the nineteenth century, and has give a brief description of it (*Industries anciennes et modernes de l'empire chinois*, p. 136, Paris, 1869).

CHAPTER II

THE HISTORY OF INK IN JAPAN

According to the *Nihongi* ('Annals of Japan'), the king of Korea sent in AD 610 two Buddhist priests to Japan, one of whom, the Korean priest Tan-cheng or Tam-ch'i from Kao-li, was skilled in preparing painters' pigments, paper, and ink. He introduced into Japan the technique of manufacturing ink and paper. This industry was ardently advocated and promoted by the celebrated Japanese prince, Shōtoku Daishi (AD 572-621), the propagator of Buddhism in Japan.

In the beginning the Japanese availed themselves only of the black derived from resinous woods like pine-tree (Japanese *sho-yen* or *matsu no kemuri*). At a later date, however, they learned from the Chinese the method of making a superior ink from the black of oil-lamps (Japanese *yu-yen* or *abura-susu*), and this process has now superseded the pine-soot method. Although Japan itself manufactures the greater part of the ink (*sumi*) required by the country, the Chinese product is looked upon as superior in quality and commands a higher price.

In principle, the process of manufacture is identical with that of China, but deviates somewhat in details. The lamps used for the purpose are small crucibles or dishes of stoneware, with wicks of rush-pitch. A cone-shaped soot catcher or reversed bowl of burnt clay, but unglazed, is placed over each lamp and is replaced with a new one every hour. The rough clay is preferred so that the black matter precipitates in the porous surface. When exposed to the flame too long, it would become too compact. The soot is carefully brushed off and swept together, and is then sifted through a fine hair-sieve.

The glue to be added to this substance is made from ox-hides and isinglass, and must be very bright, acting as it does as a cement. To ten catties (13⅓ lbs.) of lampblack from the oil of *Aleurites vernicia*, four catties of old ox-hide glue and one-half catty of old isinglass are reckoned; oil of sesame and colza also are utilized.

These ingredients, after the glue has been boiled in the necessary amount of water, are thoroughly mixed in a porcelain dish or copper basin—a toilsome process, as the lampblack does not readily combine with water. This being done, the mass may be kneaded and pressed like dough, and is shaped into round balls which are wrapped in cloth. They are placed in a stoneware jar with perforated bottom to be subjected to steam for fifteen minutes. The material is then taken out and wrought with a pestle in a mortar for at least four hours, until it is thoroughly homogeneous and plastic. It is, further, fashioned into large prismatic bars which for a moment are exposed to a temperature of about 50° Celsius in a jar, and then stretched into longer sticks. These are beaten with mallets on an anvil and constantly turned till they have acquired not only the proper form, but also the desired luster. They are once more kneaded on a smooth table; musk, camphor, or some other odoriferous substance being added, and then shaped by hand and put in a wooden press.

Ashes from rice-straw, carefully sifted and dried in the sun, are used for drying the sticks. For this purpose a layer of ashes about an inch thick is placed in the drying-box to be followed by a layer of ink-sticks; then ashes again, and another layer of sticks covered by ashes on the top. The length of the drying process depends on the quantity of water contained in the ink. When satisfactorily dry, the sticks are removed from the ashes, brushed off, laid in a small sieve, and for a day or two are left in a shady spot, where the drying process is completed. They are then polished by means of a brush, sometimes varnished and gilded, and any required legends as manufacturer's name, name of place, devices or mottoes are impressed on the surface.

This ink should not be used for several years after making, as hardness, blackness, and luster increase with age. The same rule is observed in China. The quality depends largely on the fineness and lightness of the lampblack, the purity of the glue, and carefulness observed during the several stages of manufacture. Sound and a tinge of brown color are regarded as criteria by which to recognize and judge the best pieces. Ink of the first quality is uniform, without cracks or blemish, and brilliant in its fracture. In rubbing it with water on the inkstone it must not crackle; that is, it must be entirely free from any kind of sandy matter. The odor is required to be that of a pleasing blend of musk and patchouli; the color, that of a black-brownish with a slightly russet tint. The writing when dry must have frigid and glossy tones. The sticks of prime quality, as a rule, have a plain surface without much decoration, and are completely gilded; while those of secondary or inferior quality are usually more highly ornate.

Musa from the province of Omi, Kaibara from the province of Tamba, and Taihei from the province of Yamashiro formerly were renowned brands of ink. At the present time it is the city of Nara, the ancient capital of Japan situated between Kyoto and Osaka, and the manufacturers Matsuda and Matsumura in the province of Kaga who enjoy the greatest reputation for their output of ink.

The pallets for rubbing the ink on (Japanese *sudzuri*) are made in imitation of Chinese and Korean models, usually of a fine-grained dark stone, chiefly old slate, serpentine, or colored marble. In Japan an old, dark blue slate is especially prized for this purpose, and is generally used. It is found in the neighborhood of Amabata, a small town in the province of

Kiushiu, and is hence known throughout the country as "stone of Amabata" (*Amabata-ishi*). A cavity is inserted on one side of the stone to serve as a receptacle for water. When ink is required, a few drops of water are poured into the hollow, the stick is dipped in, the water being brought up by it to the surface of the pallet. The ink-cake is rubbed against the stone, and the ink gradually flows back into the well, ready for use.

The Japanese businessman always carries with him a portable writing-case (*yatate*), including a holder for fluid ink and a writing-brush enclosed in a metal case. For household purposes is furnished a box with several compartments (*sumi-ire*)—one for brushes, another for ink-cakes, and a third for the ink-stone.

The illustration of a Japanese writing-desk may be seen in Edward S. Morse's *Japanese Homes and Their Surroundings* (p. 317). The usual form consists of a low stool not over a foot in height, with plain legs for support, sometimes having shallow drawers. This is about the only piece of furniture in a Japanese house that would parallel the style of writing-table used in the western part of the world. Paper, paper-weight, ink-stone with ink, water-bottle, brushes, and brush-rest are placed on the desk in the same manner as in China. On page 141 of the same work Morse gives a sketch illustrating the writing-place in a guest room.

In preparing ink for wood-block printing, ink-cakes are macerated in water for a few days, until the glue contained in it is dissolved and the mass becomes sufficiently softened. It is then ground by means of pestle and mortar. After it has been mixed with water, glue solution or rice paste, according to the printer's judgment, has to be added. If glue solution is employed, it should be mixed with the lampblack in a basin; but rice paste is mixed with the pigment on the plank by means of the brush.

CHAPTER III

THE HISTORY OF INK IN CENTRAL ASIA

Owing to the geographical position of the country, the culture of the country, the culture of Tibet is of a dualistic character in its absorption of foreign ideas: on the one hand these have filtered in along its southern border from India, and on the other hand along its eastern frontier from China. While the alphabet, literature, and religion were received from India, all practical industries came from China, and so it was with paper and ink.

Under the first powerful Tibetan king, Srong-btsan sgam-po, who died in AD 650, writing was introduced from northern India, and soon afterwards the king invited scholars from China to draft his official reports to the emperor, whose daughter he had received in marriage in AD 641. In 648 he applied to his imperial father-in-law for workmen capable of manufacturing paper and ink, and this request was granted. From this date onward the Tibetans joined the ranks of literary nations, and in a few centuries developed a literature of an astounding extent. They likewise adopted from the Chinese the art of block-printing, and we now have at our disposal Tibetan writings as early as the ninth century. It follows from the preceding account that the Tibetans learned the preparation of ink from the Chinese; but the bulk of their ink is still imported from China.

Colored inks, and especially writing in gold and silver, are mentioned in Tibetan literature at early date. Copying a religious book means accumulation of religious merit, and the merit is accordance with the color of the ink: gold is regarded as first in rank; silver, as second; vermilion, as third; and black, as fourth. Gold and silver manuscripts are written on a stiff, heavy paper of black glossy background surrounded by a blue border.

As early as the beginning of the fourteenth century the first copies of the Buddhist canon, known as the Kanjur and Tanjur (making about 230-245 large volumes), were produced in gold writing by Sa-skya Pandita, and subsequently we hear of a number of such editions of the sacred scriptures both in gold and silver, also in an alloy of both of these metals. A superb edition of the Kanjur in vermilion was issued in 1700 from the press of the imperial palace of Peking (the so-called 'palace edition') by order of the emperor K'ang-hsi, and another of the same character by his successor, Ch'ien-lung.

In China and Japan also Buddhistic manuscripts in gold are occasionally found; hence we may infer that this practice was propagated by the Buddhist clergy. In the West gold-writing reached its highest development among the Byzantines (cf. V. Gardthausen, *Buchwesen im Altertum*, p. 214).

The Tibetans do not rub ink on a stone, as the Chinese do, but carry it dissolved in brass ink-pots, together with a pen-case, which are suspended from the girdle. The pens are bamboo styles placed in pen cases of brass, copper, silver, or iron inlaid with silver. In the collections of Field Museum, Chicago, writing materials from Tibet, including ink, pens, specimens of paper, manuscripts and prints, and all implements used in printing, are on exhibition.

In Tibet as well as Mongolia, the pupils in the schools use as slates slabs of black painted wood, dusted over with white chalk, on the surface of which the writing is done with a style.

The Lolo, an aboriginal tribe inhabiting parts of Szechuan and Yunnan and distant kinsmen of the Chinese and Tibetans, manufacture ink from a soft schist of blood color, which is dissolved in water, and also from the ashes of a large mushroom that grows on the trunk of an oak. They use a style of a tender wood for writing, and at the present time avail themselves of Chinese paper. In ancient times they employed tree-bark for this purpose. In the collections of Field Museum are two Lolo documents written on oblong slips of wood.

In the *Mo shi* of Lu Yu, referred to above (pp. 17-18), there is a brief notice of ink in Chinese Turkestan.[10] A Buddhist monk named Su T'ai-kien is quoted as saying that in Turkestan there are neither ink-slabs nor writing-brushes (a wooden style was in use there), but only excellent ink which is not surpassed by that of China. It was prepared from old pine-trees growing in the Ki-tsu ("Chicken-foot") Mountains. T'ai-kien would keep leaves of the palmyra-palm inscribed with several hundred Sanskrit letters, the ink being twice as glossy as that of China. When at the time of the autumn rain the windows covered with such paper were wetted, the writing even though rubbed could not be wiped out.

Speaking of the documents inscribed on leather and discovered by him in Turkestan, Sir Aurel Stein (*Ancient Khotan*, Vol. I, p. 347) observes,

> "Owing to its exposed position on the outside surface, the writing of the address has often become faint or been partly rubbed off. But the ink on the obverse has in most cases retained remarkably well its original black color, and makes the writing clearly legible even in those cases where the leather itself has become discolored or stained. I regret not to have found an opportunity for arranging for a chemical examination of this

---

[10] Now Xinjiang Province, China.

ancient ink. But, judging from its appearance, it seems probable that it was Chinese (or Indian) ink, such as that of which a small stick was actually found by me among the rubbish layer inside the Ender Fort. The ink used on the tablets, both Kharoshthi and Chinese, varies considerably in quality and thickness, but I did not observe any indication pointing to a difference in the composition of the ink."

The Mongols appear to have borrowed their ink from China at a comparatively early date, as is proved by their word *bäkhä*, which is based on Old Chinese *mak*, *bak*, or *bäk*. The Manchu adopted the same word, presumably from the Mongols. In this connection it is interesting to note also that Mongol *sir* ("varnish") is borrowed from Old Chinese *tsit* or *tsir*, and Mongol *bir* ("writing-brush") from Old Chinese *bir*, *bit*.

In 1848 the great Finnish linguist, A. Castrén, wrote from Kiachta, "In the art of printing, the Mongol Lamas are comparatively less skilled than in writing; but it is curious enough that this very art is practiced in this barbarous country. The Lamas, in accordance with their regulations, are obliged to know how to cut printing-blocks, to prepare printer's ink, and to print from the blocks."

CHAPTER IV

THE HISTORY OF INK IN INDIA

In considering the history of writing materials in India we are at the outset confronted with a psychological situation radically distinct from in China and even almost the opposite to it. In China the written word and everything connected therewith were regarded with fervent reverence and treated as a fetish. Among the Brahmans of ancient India, it was not the written, but the spoken word which was looked upon as a fetish. The hymns of the Veda were memorized and transmitted for ages from generation to generation merely by memory; even at a time when an alphabet was in existence, the Brahmans first steadfastly refused to commit their sacred texts to writing, and but slowly and reluctantly yielded to this far-reaching innovation which threatened to break down the prerogatives of their caste. China never labored under a caste system; China has always been democratic in this respect, and placed the means of learning and education in the hands of whoever endeavored to learn and to read.

In India learning was the privilege of an exclusive sacerdotal class which kept in splendid isolation. In the *Mahābhārata* it is said that those who sell, forge, and write the Veda are condemned to hell. In a society where such an aversion to writing prevailed it is not likely that much interest was evinced in the production and perfection of writing materials. It is striking also that despite her close contact with China, which set in from the first century AD, India did not adopt paper and printing. Paper was introduced only in the Mohammedan period by the Arabs (there is no Sanskrit word to designate paper), and the first printing press in India was set up by the Portuguese at Goa in the sixteenth century. Whatever progress was made in India in the direction of writing must have been due to the caste of nobles and warriors, the Kshatriya, and to the merchants.

In the sixth century BC there were schools for methodically teaching the art of writing. Wooden writing-boards (*phalaka*) were in use, but these were incised with a stylus. The whole terminology relating to script and scribes hints at the fact that the letters were scratched in hard objects. There in no vestige of the use of ink in the early period. In the fourth century BC we learn from the Greek writers that prepared cotton-stuffs and birch-bark were employed in India, like papyrus, for writing letters. From this fact G. Bühler (*Indische Palæographie*, p. 91) is inclined to infer that ink was presumably used, and he confirms his supposition by palæographic evidence.

From the second century BC we have the oldest extant specimen of ink-writing on a stone vessel recovered from the tope (stupa) of Andher. In post-Christian times we have manuscripts written on birch-bark, the oldest being the small leaves folded and fastened with yarn (so-called "twists") discovered by Masson in the stupas of Afghanistan,[11] followed by the famous Bower Manuscript which goes back to the fourth century AD. Hoernle, in his edition and translation of the Bower Manuscript, says nothing concerning the ink. Aside from birch-bark, the leaves of several species of palm were enlisted as writing material in early times. Hüan Tsang, the famed Chinese Buddhist pilgrim, who visited India in 629-645, states that the leaves of the *tala* palm, which are

---

[11] See G. Whitteridge, *Charles Masson of Afghanistan*, Bangkok 2002, p. 104.

long and broad and bright in color, are everywhere used for writing on in all countries of India.

In the Horiuji Monastery of Japan is preserved a Buddhist palm-leaf manuscript inscribed with ink, and numerous such manuscripts of the ninth and later centuries from Nepal, Bengal, Rajputana, Gujarat, and the northern Deccan demonstrate that in northern, eastern, central, and western India ink was used in writing upon palm-leaves. In Orissa and Dravidian India, however, the letters are incised in the leaf with a metal style, and are subsequently blackened with soot or charcoal. The use of palm-leaves for manuscripts is still common in southern India; the oldest manuscript extant there comes down from AD 1428. Palm-leaves were also used in southern India for letters, as well as official and private documents, and are still so used; there, and in Bengal likewise, they serve for writing in school. In the schools of Bengal banana-leaves also are said to be used and inscribed with lampblack ink.

In the *Vāsavadattā*, a Sanskrit romance written by Subandhu in the seventh century AD and translated into English by L. H. Gray, occurs the passage, "The pain that has been felt by this maiden for thy sake might be written or told in some wise or in some way in many thousands of ages if the sky became palm-leaves, the sea an ink-well (*melāmandā*), the scribe Brahma, and the narrator the Lord of Serpent."

John Fryer, who traveled in India and Persia for nine years (1672-81), informs that the Persians "use Indian ink, being a middling sort betwixt our common ink and that made use of in printing: instead of a pen they make use of a reed, as in India."

We have a well-authenticated testimony for the existence of ink in India in the first century of our era in the *Periplus of the Erythrean Sea*, a Greek work from the hand of an unknown author, probably written between AD 80-89, roughly about AD 85. In chapter 39 of this book, *Indikon melan* ("Indian black") is given as one of the articles exported from the Indian port Barbarikon. The earlier commentators have explained this term as indigo, and B. Fabricius (in his edition of the *Periplus*, p.152) is even inclined to interpret it as textiles made in India and dyed black. These opinions are not to the point, for the indigo of India is called in Greek simply *indikon*, in Latin *indicum* (cf., further, *Sino-Iranica*, pp. 370-371), while *melan* is the common Greek designation for ink: *Indikon melan*, consequently, means "India ink. Moreover, Pliny (*Hist. nat.*, XXXV, 25), in his chapter on ink (*atramentum*), points out *"indicum*, a substance imported from India, the composition of which is at present unknown to me," and says expressly that good ink prepared from dried wine-lees will bear comparison with that of *indicum*. Indigo is discussed by Pliny in a separate chapter.

There is hence no doubt that *indicum* signifies "ink of India," which was exported from India into the Roman Empire, and as confirmed by the *Periplus*, shipped together with other Indian goods from Barbarikon. Old Beckmann (*Geschichte der Erfindungen*, Vol. IV, 1799, pp. 489-496; cf. also Blümner, *Technologie*, Vol. IV, p. 517) has devoted a profound and ingenious investigation to the whole question, and has arrived at the same result. He thinks it very probable that the "Indian black" of the ancients was nothing but what is now termed India ink which approaches the finest ivory-black and lees-black so closely that by this means some still imitate it and actually delude ignorant buyers. "Ink in India," he concludes, "is in general use, and has presumably been so from earliest times; for in India almost all products of art are extremely ancient, but I do not mean to say that ink is a new

Indian invention; it may have been improved, above all, by the Chinese."

There is no evidence to the effect that the Indian ink of the ancients was Chinese ink. In the first century AD, as we have seen, it was still in its initial stages and very far from the perfection of the later days. All that we are permitted to assert safely is that the Indian ink of the ancients was an ink manufactured in India.

Blümner argues that the manufacture of Chinese ink is exceedingly old, and that in the same manner as Chinese silk was traded to the West, also ink might have arrived in Europe by way of India. Its native country being unknown, it was designated as Indian. There is, however, not a trace of documentary evidence for such a trade in ink, either in Chinese or in Western sources; and Blümner also adds cautiously that Chinese ink has not yet been traced in any paintings or pigments of classical antiquity; all investigations of black pigments have only yielded substances consisting of pure carbon.

The oldest Sanskrit designation for ink is *maṣi* or *maṣī*. The word is indigenous, and according to Bühler, originally means "something ground, powder." It then came to denote several kinds of powdered charcoal which was mixed with gum-arabic, water, and sugar, and thus served as an ink. Another name for ink, *melā*, has been derived by some scholars from Greek *mélas* ("black"), but Bühler rejects this view and connects the word with Prakrit *maila* ("dirty, black"). According to L. D. Barnett (*Antiquities of India*, p. 231), ink was made in early times of charcoal mixed with water, sugar, gum-arabic, etc., and was applied with pens of wood or reed. A solution of chalk was also used as writing fluid, and was conveyed to the tablet by a wooden style.

In modern times the ink used for writing on paper is compounded of lampblack with an infusion of roasted rice, with the addition of a little sugar and sometimes the juice of a plant called *kesurte* (*Verbesina scandens*). It requires several days' continued trituration in a mortar before the lampblack can be thoroughly mixed with the rice infusion, and want of sufficient trituration causes the lampblack to settle down in a paste, leaving the infusion on top unfit for writing. Occasionally, acacia gum is added to give a gloss to the ink; but this practice is not common, sugar being held sufficient for the purpose. Of late, an infusion of the emblic myrobalan, prepared in an iron pot, has occasionally been added to the compound; but the tannate and gallate of iron formed in the course of preparing this infusion are injurious to the texture of paper, and Persian manuscripts sometimes written with such ink suffer much from the chemical action of the metallic salts.

The ink for palm-leaf consists of the juice of *Verbesina scandens* and a decoction of *alta* (cotton impregnated with lac dye). It is highly esteemed, as it sinks into the substance of the leaf and cannot be washed off. Both these inks are very lasting, and being perfectly free from mineral substances and strong acids, do not in any way injure the paper or leaf. They never fade and retain their gloss for centuries (after A.E. Gough, *Papers rel. to the Collection and Preservation of the Records of Ancient Sanskrit Literature in India*, p. 18, Calcutta, 1878).

Colored inks with which especially the Jaina produced beautiful manuscripts are frequently mentioned in Brahmanic literature, e.g., in the *Purāṇa*s when donations of manuscripts are mentioned. Chalk and minium served as substitutes for ink in ancient times.

According to G. Watt's *Dictionary of the Economic Products of India*, at present various substances are used by the natives of India in marking ink, the usual process being to mix some astringent principle such as galls or myrobalans with one of the iron salts or oxides. In Madras charcoal of the rice plant is employed in combination with lac and gum-arabic, and the Mohammedans generally prepare their ink from lampblack, gum-arabic, and the juice of the aloe. The following are the plants specially mentioned as adjuncts in the formation of ink:

(1) *Alnus nepalensis*, D. Don. Bark forms an ingredient in native red inks.

(2) *Cordia myxa* L. The unripe fruit is said to be used as a marking ink, though its color is less enduring than that from *Semecarpus*.

(3) *Phyllanthus emblica* L (Sanskrit *amaleka*). Fruits are largely employed in making black ink.

(4) *Semecarapus anacardium* L. (cf. *Sino-Iranica*, p. 482). The marking-nut tree bears a fruit with fleshy receptacle which contains a bitter and astringent substance universally used in India as a marking-ink, the juice being mixed with lime water as a mordant. Without the addition of lime it is often employed as ordinary writing-ink. As it is apt to cause severe inflammation, it has to be used with caution.

(5) *Terminalia belerica* and *T. chebula*, the unripe fruit of either species, or indeed of any *Terminalia*, is combined with iron in making ink.

The Siamese largely make use of Chinese ink, with which they write on long strips of gray paper made from tree-bast. A professional class of writers, called *alak*, avails itself ordinarily of a gum-resin dissolved in water, writing in yellow script on black paper. The Buddhist scriptures of the Siamese composed in Pali are written in Indian fashion on palm-leaves, the characters being incised by means of a style.

In ancient Cambodia, according to the account of Chou Ta-kwan, who visited the country in the thirteenth century, official and private documents were written on pieces of deer-skin dyed black. They availed themselves for writing of a white clay, probably chalk, resembling the *kaolin* of China, moulding it into sticks, which were handled like pencils. Paper and ink were introduced from China and were used at an early date.

CHAPTER V

PAPER AND PRINTING IN CHINA
AND KOREA

The ancient Sumerians, Babylonians, Egyptians, and Greeks may have reached a flourishing civilization long before the Chinese, but all their achievements, however great, do not equal in importance the invention of paper which we owe to the Chinese and the art of printing that was born from it. Printing has been, and still is, the supreme factor in the progress of civilization. The Chinese as the inventors of paper were the first who printed books, many centuries before Gutenberg, and not only that—they have also made typography a fine art and produced books which belong to the finest examples of the craft.[12] They have been a book-loving people for ages. The primary conditions of printing are paper, writing-brush, ink, and ink-pallet or ink-stone, which are still regarded by the Chinese as the four great emblems of scholarship and culture that form the fundamentals of their civilization. These four constituents the Chinese may justly claim as their own, inventing them and perfecting them entirely from their own resources, unaided

---

[12] This holds true down to modern times: the great Swiss typographer and graphic design artist, Jan Tschichold, declared the 1952 edition of the *Shi Zhi Zhai Jian Pu* ["Collection of Letter Paper from the Ten Bamboo Studio"], woodblock-printed in Beijing by the Rongbaozhai studio, "…incomparably, perfect…the best printed book of modern times anywhere."

by any other nation; and this arsenal has largely contributed to make them a nation of learned, studious, well-bred and cultured people.

The turning point in the history of printing is the invention of paper by Ts'ai Lun in AD 105. In order to understand and appreciate this event correctly, it is necessary to have some idea of what writing-materials were, prior to that date, and in what condition the early documents and books were before the art of printing came into being. I shall therefore discuss, first, Chinese books before the invention of paper and, second, Chinese books after the invention.

The earliest means of communication in ancient China of which we have any knowledge reminds us of the quippus of the ancient Peruvians.

In a prehistoric age we find in China knotted cords in use for the conveyance of messages, chiefly in the transaction of government business. Lao-tse, the famed philosopher, in a sentimental yearning for the past, expressed the desire that he might bring his people back to the ancient usage of knotted cord; that is, the simple life of old. The Tibetans have a tradition to the same effect, and several aboriginal tribes in the south of China availed themselves of this method as late as the twelfth century of our era.

In early historic times, calendars, calculations, and contracts were made by means of wooden tallies in which notches were carved with a knife. Even when writing had long been in use, contracts made by means of notches in a wooden stick were continued for simple business transactions—such as deeds, bonds, or obligations. The wooden stick was notched on either side and then split and equally divided between the two contracting parties, the creditor receiving the left half, the debtor the right half of the tally. When the time arrived for fulfilling the contract, the two halves were joined together to make sure that the notches of the one tallied with those of the other. When the debt was liquidated, all the creditor had to do was to break up his portion of the wooden contract. This was called "breaking the contract," which meant as much as "fulfilling one's obligation." Credit systems were always highly developed in China, and there is an old story on record that a tailor even made garments on credit for a duke and that whenever the duke was in a position to render a payment on the instalment plan, the wooden tally was smashed by the happy tailor. Even when contracts were subsequently drawn up in writing, the notches were retained as a means of checking the two halves or verifying the twin documents.

A survival of this practice still characterizes the modern banking system. Our banking methods are based of the signature and identification of the individual. Neither is required in China. A Chinese draft is made out in duplicate on a single long sheet of paper, containing the same matter on the right and left sides, one column of writing running exactly down the center. The document is evenly cut into halves across this line, the right half being given the bearer of the draft, the other half being mailed to the bank on which the draft is issued. When the bearer presents his half of the draft at this bank, it is carefully checked off and tallied with the other half which meanwhile arrived at the bank by mail; and if the two halves are found to fit perfectly together, payment is made, no receipt and signature being required.

These drafts bear a striking resemblance to the indentures used in old England. Hamlet, in musing over a lawyer's skull, exclaims, "Will his voucher vouch him no more of his purchases, and double ones too, than the length and breadth of a pair of indentures?" They were referred to as a pair, as both copies of a deed were written on one piece of parchment or paper and then cut asunder in a serrated or sinuous line (a reminiscence of the notches in tallies), so that when brought together again the two edges exactly tallied and proved that they formed part of the same document.

A fundamental of culture in eastern Asia is divination. Divination was based on the bones of certain animals. In central Asia, divination was practiced from the shoulder-blade of a sheep which was scorched over a fire, and from the cracks thus arising in the bone the future was predicted. In ancient China the carapace of a tortoise was utilized in fortune-telling, and this magical procedure probably gave the impetus to the origin of writing. The tortoise was regarded as a sacred animal imbued with a knowledge of the future. In 1899 a deposit of several thousand fragments of bones, chiefly tortoise-shell, was discovered at Chang-te fu, Honan Province. These bones, forming a sort of archive, are engraved with inscriptions of a very archaic style, representing the earliest form of Chinese script we now possess, and were used for purposes of divination. They date in general to about 1500 BC. The oracles and in some cases the answers were incised into these bones. We meet, for instance, inscriptions such as these: "We consulted the oracle to ascertain whether the harvest will be abundant." or "The oracle was consulted, as we wish to know whether God will order a sufficient rainfall so that we may obtain an adequate food supply," or "If we go hunting tomorrow, shall we capture any game?" Divination has always dominated the whole life of the Chinese from the cradle to the coffin, and no business was transacted, no marriage concluded, no burial undertaken, without consulting a fortune-teller. These ancient augurs formed a special profession, in their social position comparable to the lawyer of our society. In the same manner as the modern financier and captain of industry consults his lawyer on all important questions, so the Chinese did not make a move in the most trivial matters without asking a diviner's advice.

Further, we have from the early dynasties inscriptions on objects of bronze such as vases, bowls, bells and weapons, cast by means of the lost-wax process, the characters being traced in the wax mould, and being either incised or raised in the bronze. Tablets of jade were used for writing by the emperors; tablets of ivory, by the nobles and higher officials. The most common writing-material, however, particularly under the Chou dynasty (1122-247 BC), consisted of bamboo slips or square wooden splints which were perforated at their upper ends and fastened together by means of a silk cord or fine leather strip. The main difference between the utilization of bamboo and wood was this, that a message containing upwards of a hundred words was written on bamboo slips; when it contained less than a hundred words, on wooden boards. The bamboo tablets were naturally narrow, and could be piled up in any required number, formed into a pack, The wooden documents, being too heavy to allow of a combination of many, served only for brief texts, as official acts and regulations, statistics of the population, and prayers, but they could not be united into books.

The early canonical or classical literature was handed down on bamboo slips of different lengths, each slip as a rule containing a single line of writing varying from eight to twenty-five or thirty words, and inscribed on one side only. A great number of such tablets was naturally required to make a book. Such books, of course, were exposed to many causes of destruction, chiefly from humidity and pernicious insects, so that bamboo books of early antiquity have long since disappeared, but a large number of wooden documents of the Han period have come to light in Chinese Turkestan.

Another inconvenience of these books was their heavy weight. A curious incident in allusion to this fact is recorded concerning the emperor Ts'in Shi who was compelled to examine daily state documents to the weight of a hundred and twenty pounds.

In 501 BC the penal code of the Chou dynasty was promulgated on bamboo slips. The edicts of the Han emperors were inscribed on bamboo tablets, alternately a long and a short one; one two feet, the other one foot long, each pair fastened together by cords of undyed silk. The peculiar condition of the ancient manuscripts is, of course, an important factor in the critical study of ancient literature, as it frequently happened that one or the other tablet became lost or that, if the band of silk or leather broke, the tablets became mixed up, and it was a difficult task of subsequent editors to restore them to order. In some ancient books entire chapters are now lost because the pack of bamboo slips suffered destruction.

In the third century before our era two novel writing-materials came into use—pure silk in the form of bands of silk stuff and a sort of silk paper made from silk refuse including both raw and woven silk. The refuse from silkworm cocoons was soaked and beaten in water in order to eliminate coarse particles. This mass was reduced to a paste and purified, and then was spread over a fine bamboo mat mounted on a wooden frame. The latter served as a mould on which the paste precipitated, and when dried, produced a sheet of paper. This was a sort of "near" paper consisting of animal fibers. It is obvious that the principle underlying the subsequent manufacture of rag-paper was forestalled by that of silk paper. This new material resulted in the perfection of writing-brushes of pliable hair, for the ancient wooden stylus could hardly be applied to silk.

It took the Chinese several centuries of tests and trials until they discovered an acceptable formula for a good ink. Under the Han dynasty ink was prepared from mineral products, such as graphite, mineral coal, bitumen, and rock-oil. From the third century AD onward ink was made from vegetal matters, the chief ingredient being lamp-black obtained from pine wood, that was blended and boiled together with glue and aromatic substances. The glue serves the purpose of uniting the fine particles of carbon and fixing the ink on paper by means of the brush. The perfumes sometimes added, such as musk, camphor or patchouli, have the function of camouflaging the unpleasant odor of glue, but are unessential. (See Ch. I above.)

Paper was invented in China by Ts'ai Lun in AD 105 when he conceived the idea of manufacturing with refuse material of vegetable origin a substance light and economical at the same time, which would replace advantageously the writing-materials used up to his time. A record of this memorable event is contained in the biography of Ts'ai Lun, which is embodied in the Annals of the Later Han dynasty (chap. 108). Ts'ai Lun was born at Kweiyang, a city in Kweichou Province in southern China. In AD 75 he entered the service of the Emperor Ho, and in AD 89 was appointed director of the imperial arsenals. He was deeply given to study, and whenever he was off duty, he would shut himself up for that purpose. The passage relating his memorable discovery runs as follows:

> "From times of old, documents had been written on bamboo boards fastened together. There was also paper made of silk refuse (*chi*). But silk was too expensive, and the bamboo boards were too heavy; both were inconvenient. Therefore Ts'ai Lun conceived the idea of utilizing tree-bark or bast-fiber, hemp, and also old rags and fishing-nets for making paper. In AD 105

he submitted his invention to the emperor, who lauded his skill. From this moment there was no one who did not use his paper, and throughout the empire, all people called it the 'paper of the honorable Ts'ai.'"

This brief and sober account reveals what the writing-materials were in the times before Ts'ai Lun, what his innovation consisted of, and what impression it made on his contemporaries. It should not be understood that the ingredients enumerated in this passage were mixed together and resulted in paper; but each substance was the principal constituent to make a particular kind of paper. Paper may be obtained from many and various plant-fibers by a process of cleaning, maceration, and drying, The paper of Ts'ai Lun was in fact distinguished, according to the material used, as "hemp paper," "bark paper," etc. He substituted vegetal fibers for the fibers of animal origin used previously; in principle he utilized two kinds of materials—the raw fibers of bark and hemp and the worked-up fibers of rags such as ropes and fishing-nets. He survived his invention for thirteen years, and was ennobled in AD 114 as marquis by the empress dowager. He was no favorite, however, with the empress; and when his patroness, the empress dowager, died, the empress began to intrigue against him. Driven to despair, he died a suicide by swallowing a dose of poison.

Two different places were pointed out in later times as the seat of Ts'ai's operations. According to one report, he had lived near Leiyang in Hunan Province, where near his residence was shown a stone mortar in which it was said he had pounded his paper pulp. Under the T'ang (AD 618-906), this mortar was sent by the city as a gift to the imperial court. According to another tradition, his residence is claimed by the city of Tsaoyang in Hupeh Province, where nearby there was a pool, called Ts'ai's pool; and there it is said he first manufactured paper from fishing-nets. His art, a Chinese chronicle relates, is hereditary among the people of that district, many of whom are still expert at the manufacture of paper.

Ts'ai Lun's invention created a profound impression on his contemporaries. Although he perfected a pre-existing process and his principal merit consisted in substituting but little valued or even valueless substance, which yielded better results, for the comparatively costly silk-refuse, or in other word, substituting vegetal fibers for animal fibers, he merits high honors as the man to whom we are indebted for one of the most far-reaching discoveries ever made in the annals of technology. Without paper there would be no adequate record of the past, no history, no science, no progress. The manufacture of paper denotes a landmark in the intellectual development of mankind; it sets off civilization from the stage of savagery.

Archaeological discoveries in Chinese Turkestan, where the dry sand has preserved numerous old manuscripts, have signally confirmed the correctness of the Chinese account of paper invention. We now have thousands of specimens of Chinese rag-paper from Turkestan datable as early as AD 150 and going down to the ninth century. Many chemical and microscopical analyses of these ancient paper remains have been made, especially by J. Wiesner of Vienna, who found that mulberry bark and ramie or China grass (*Boehmeria nivea*) were the most common materials in the paper manufactured in northern China.

The manufacture of paper remained a Chinese monopoly until the year AD 751 when, after a battle with the Arabs, Chinese captives, who were familiar with the technique of making paper, introduced it into Samarkand. From that date onward it became a permanent acquisition for the

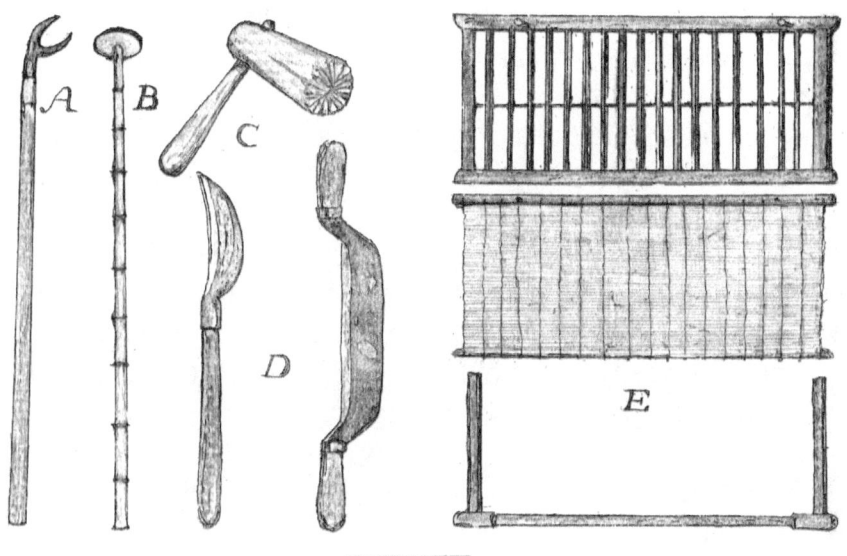

EXPLANATION.

A: An iron hook for dredging up the bamboo.
B: A bamboo pole used to stir the pulp in the vat.
C: A mallet for smashing the bamboo.
D: Bamboo cutting knives.
E: [top] The wooden frame of the mould.
[center] The bamboo screen which lies on the frame.
[bottom] The deckle which fits on the screen.

Fig. 7. Bamboo Papermaking Tools, after *Out of the Mountains,* by Yan Ruyi, 1822; privately printed, Honolulu 1991. The earliest Chinese papers typically were fabricated of hemp or mulberry bark fibers, but by the eleventh century bamboo—one of the most widely available plants in the country—became popular as a raw material for paper construction.

Mohammedan world. Under the reign of the Caliph Harun ar-Rashid, the industry spread to Baghdad in AD 794, and paper factories soon sprang up in all Islamic countries—in Persia, Arabia, Syria, Egypt, and Spain. Damascus paper attained a special celebrity and was largely imported to Europe. By the middle of the tenth century the use of papyrus was superseded by paper throughout the Arab dominion. In AD 1035, it is reported, the grocers of Cairo used paper for wrapping the goods sold to customers, which means that it must have been a common article at that time. About AD 1150 a paper-mill was founded at Fabriano, Italy, which became the principal center for paper-making, and this region continues the manufacture to the present day. From Italy the art spread to France and Germany, somewhat later to England, where it began to flourish when the Revocation of the Edict of Nantes in 1685 sent many French paper-makers into exile to England and America. In 1690 (a millennium and a half after the first invention) the first paper-mill was organized in America by Willam Rittenhouse at Roxborough, near Philadelphia.

During the early centuries of our era, paper of a great variety, paper sized and loaded to improve its quality for writing, paper of various colors, writing paper, wrapping paper, even paper napkins and toilet paper, all were in general use in China. The method of sizing paper with starch is also an improvement initiated by the Chinese. Their ancient practice of extracting the fiber from the bark and other parts of plants by means of maceration is in principle identical with our modern method of extracting cellulose by means of chemical processes.

In the tenth century the Chinese conceived the idea of issuing printed paper money, which reached its climax under the Mongol emperors. It aroused the greatest admiration of Marco Polo, who devotes to it one of the most interesting

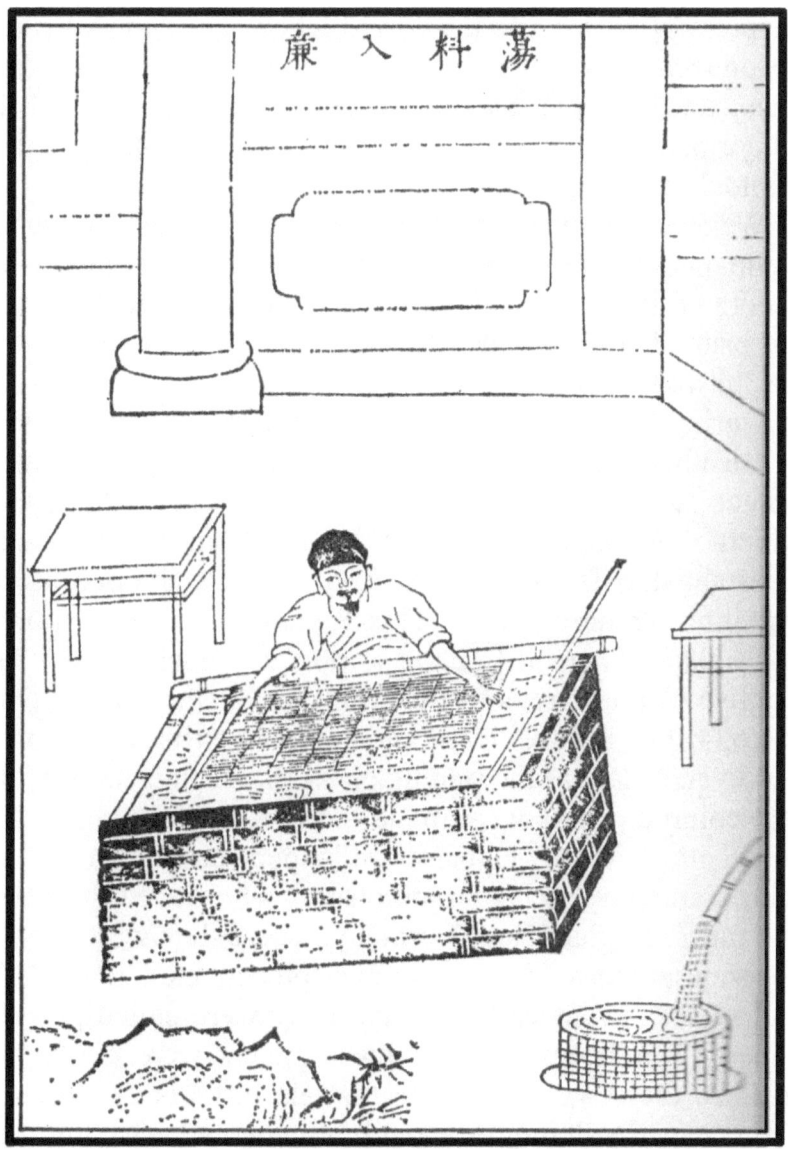

Fig. 8. Collecting bamboo pulp on top of a screen, from an old xylographic print reproduced in *Chinese Technology in the Seventeenth Century*, Sung Ying-hsing, Pennsylvania State University Press, 1966.

chapters of his visit to the Grand Khan. The Mongol rulers introduced paper-bills into Persia under the Chinese name *chao*, and in 1293 established a printing-office at Tabriz, where paper-notes were turned out by the Chinese method of block-printing.

We owe to China in particular also our paper-hangings or wallpaper. The walls and ceilings of rooms are invariably decorated in China with paper, on which different patterns are printed from wooden blocks. The paper is confined in size to foot-square sheets.

During the middle ages, Europe had only linen, silk, and leather tapestry. French missionaries in China sent to France some specimens of colored Chinese wallpaper, which stimulated a Frenchman, Le François by name, to establish a factory at Rouen in 1630 for the purpose of imitating Chinese papers. This Rouen paper was exported to England where it became known as "flock-paper." The English claim a previous invention by Jeremy Lanyer who, in 1634, had used Chinese and Japanese processes. It was, however, as late as the middle of the eighteenth century that real colored papers were made in France and England. The actual importers of Chinese wallpapers painted or printed with pictures were Dutch merchants, who traded them also to France, England, and Germany, where they were used to decorate screens, desks, chimney-pieces, etc., toward the end of the seventeenth century. They were called pagoda-papers. The prices paid for these papers were exorbitant. In 1770 there was advertised for sale in Paris 24 sheets of China paper with figures and gilt ornaments 10 feet high and 3½ feet wide, at 24 *livres*[13] apiece, to be sold altogether, or in lots of 8 sheets each. By that time entire rooms were papered.

---

[13] Approximately $250 US Dollars in 2019 currency.

In 1779 an apartment in Paris was advertised to let, having a pretty boudoir with China paper of 13 sheets in small figures representing arts and crafts.

Wallpaper was brought to America in 1735. Its manufacture was introduced into the United States in 1790 by two Frenchmen, Bouler and Charden, and only three or four firms engaged in the business before 1844. In that year the first machine for printing wallpaper was put up in the Howell factory at Philadelphia. Up to that date, wallpaper had been made in small sheets of 22 by 32 inches, according to the Chinese fashion. After the establishment of machinery, continuous rolls or webs of paper came into general vogue. The dependence on Chinese models is illustrated also in the processes hitherto applied to wallpaper. The patterns were either put in with stencils and the background with a brush, or by means of block-printing, the design being engraved on a wooden block.

In old colonial mansions of Massachusetts, specimens of Chinese wallpaper are still to be found, some even imported in AD 1750 and still in a good state of preservation. Many of the older American papers exhibit their relationship to the Chinese in that the decoration is not repeated, but runs continuous about the entire room or contains a scenic representation. An interesting book on this subject was written by Kate Sanborn, *Old time Wall Papers, an account of the pictorial papers on our forefathers' walls* (Greenwich, Conn., 1905).

When the invention of rag-paper was made, the Chinese were in the possession of all technical materials that make the primary conditions for printing: an extensive literature; a suitable and economic medium, easily manufactured in large quantity, for taking print; and ink as the medium to fix permanently to the paper the written thought. And yet, it is astounding that centuries elapsed before any steps were taken in the direction of printing. This is the more amazing, as printing of an embryonic type was practiced long ago by means of seals made of clay or metal in which the script stands out in the negative in the same manner as in the later block-prints. The case is of psychological interest inasmuch as it shows that new inventions depend not merely on the existence of mechanical appliances, but to a much higher degree on the mental attitude of society. Some dynamic force is required to set the slumbering spark afire, in order to create the demand for printing. The art of printing is the first step toward democracy, the education of people, and national awakening. To this the intellectual minority in all countries of Asia and Europe was at first bitterly opposed. In ancient Egypt, if the idea of printing had ever been proposed, it would at once have been nipped in the bud by the caste of jealous priests. A similar situation prevailed among the Brahmans of ancient India, where the sacred hymns of the Veda were memorized and transmitted for ages from generation to generation merely by memory. Even at a time when an alphabet was introduced, the Brahmans first refused steadfastly to commit their sacred texts to writing, and but slowly and reluctantly yielded to this innovation which threatened to break down their monopoly and the prerogatives of their caste. In India, in opposition to China, it was the spoken word which was looked upon as a fetish. In China, it was the written word that was regarded with fervent reverence and treated as a fetish. This worship of the written word ultimately led to its permanent preservation in print, while in India this idea was always detested. Despite her close contact with China, India did not adopt from her paper and printing. Paper was introduced into India only in the Mohammedan period by the Arabs after the twelfth century, and only as late as the sixteenth century was the first printing press set up by the Portuguese at Goa. The first book printed there was Garcia da Orta's *Coloquios dos simples, e drogas he cousas mediçinais da India,* ("Colloquies on the Drugs of India,") 1563.

We might expect that printing should have arisen in the circle of Confucian scholars who certainly had an interest in preserving and diffusing their philosophical writings; but this expectation is not fulfilled, perhaps for the reason that the Confucian doctrines appealed largely to the intellectual minority, not to the masses. The people found more satisfaction in the tangible teachings of Buddhism which promised them speedy salvation in the paradise of the Amitābha. The earliest attempts at printing from wooden blocks were therefore made in the camp of the Buddhist priests and consisted of charms, especially for the healing of diseases, prayer formulas and engravings of religious images, made with the avowed object of appealing to the sentiments of the people. The earliest of these charms extant were printed in Japan about AD 770 by order of the empress Shōtoku, in fulfillment of a vow: one million charms were printed, placed in a million small wooden pagoda models and distributed among the populace, these charms being believed to be efficient in expelling the demons of disease. Several of these are still preserved. The first object of printing, therefore, was not, as we might imagine, a desire for the diffusion of knowledge, but a desire on the part of an empress to acquire religious merit and to safeguard her people from the ravages of epidemics. The meaning of the charms was even unintelligible to the people, for they were written in Sanskrit, transcribed in Chinese characters.

It is uncertain to what date block-printing in China goes back. While no accurate date can be fixed and while there is no record of an inventor of block-printing, which was achieved by a gradual process, there are indications that the initial stages are traceable to the sixth and seventh centuries. During the ninth century printing from wooden blocks was practiced in the farthest west of China, the province of Szechuan, which seems to be the home of the art. Again, the books printed there were not intended for scholars, but for the people. They consisted chiefly of works on divination, dreams, geomancy, and elementary school books, but, as our Chinese informant writes, they were so spotty and blotted that they were difficult to read.

The earliest printed book in existence was discovered in the library of the cave temples of Tun-huang by Sir Aurel Stein in 1907, and is now in the British Museum. It is the Chinese version of a Buddhist Sūtra, the *Vajracchedikā*.[14] It bears the date AD 868, contained in the colophon at the end. It reads, "Printed on May 11, 868, by Wang Chie, for free general distribution, that the memory of his parents be reverently perpetuated." In this case the printer performed an act of filial piety. The text is printed on a roll of paper 16 feet long and 1 foot wide; it consists of seven sheets pasted together, and was printed from seven blocks. The frontispiece is the earliest datable woodcut.[15] Printed books and single sheets, however, were exceedingly rare in the temple library of Tun-huang; among thousands of manuscript rolls only four printed rolls were found and a not inconsiderable number of charms

---

[14] See Ch. I, p. 22, fn 2.

[15] Despite the fact that the *Vajracchedikā* is no longer the oldest surviving woodblock printed text, the earlier text found in Korea did not include any pictorial elements. Hence the *Vajracchedikā* frontispiece remains the oldest extant woodcut. See Fig. 1 above.

from single-page block-prints. The latter presumably were of local manufacture, the former imported from Szechuan. Printing was practiced under the T'ang (AD 618-906) to a limited extent only, and did not supersede the manuscripts which were evidently regarded as more meritorious.

A certain official, Feng Tao (AD 881-954), is credited by some with the invention of block-printing. He was a versatile politician who served under no less than ten emperors of four different houses. Presenting himself at the court of the second emperor of the Liao dynasty, he asked for a post. He said he had no home, no money, and very little brains—a statement which appears to have recommended him strongly to the sovereign, who at once appointed him grand tutor to the heir-apparent. Block-printing was certainly known long before Feng Tao's time, but he was the first who applied it to the printing of the Confucian classics, and this is the reason why Confucian scholars have stamped him as the inventor of block-printing. Under Feng Tao's direction the ancient canonical literature with its host of commentaries was printed for the first time in 130 volumes. The work of editing and printing lasted for twenty-one years and was completed in AD 953. Up to the year 1064 the private printing of the classics was forbidden. All printing was the prerogative of the government and had to contain the orthodox accepted text. Of Feng Tao's edition nothing unfortunately has come down to us. The great renaissance of culture that took place under the Sung dynasty (AD 960-1279) resulted in an enormous output of literature and a corresponding advance in the art of printing.

As to the technique of the block-print process, it is simplicity itself: the book undergoes only three principal stages. What is composition among us is performed by a professional calligrapher, who receives the manuscript from the author's hands and writes it out in clear and uniform style on thin sheets of paper. Prefaces contributed by friends (and most such books are introduced by a number of prefaces) are usually facsimile from their own hand-writings, and may even be written in an archaic or highly ornamental style. Calligraphy, like drawing and painting, is an art, and the three are closely interrelated. A point worthy of note is that the Chinese scribe, as well as the draftsman and painter, is deprived of the privilege of making correction or alterations. Chinese paper and silk are highly absorbent materials, and a stroke of the brush, once made, will stand forever, and cannot be erased. The artist therefore must be sure of a firm hand and a scrupulously thoughtful and precise technique.

When the manuscript is completed, it is sent to the block-engraver. The single sheets are pasted over the finely planed and smooth wooden blocks, usually of pear-tree wood, the writing turned face downward. As the paper is thin and transparent, the writing is perfectly displayed through the back. Then commences the task of the engraver who, with a set of gouges, picks, and chisels, pares the surface of the block around the characters, so that the script in negative stands out in a flat relief.

In this state the blocks are finally transmitted to the printer whose requirements are limited to just two brushes. He uses a round, bell-shaped brush of coir-palm fibers for rubbing ink over the block. Then he places a sheet of paper over it and takes the impression by means of a flat, handled brush, which takes the place of our press. The printed sheet, of course, represents an exact facsimile of the original manuscript, and the printer cannot make any mistakes. A single sheet, as a rule, consists of two pages with a margin in the center, that contains the title of the book on top, chapter and folio number in the middle and usually title of the particular chapter at the foot. These sheets are folded and then stitched at the ends.

The Field Museum has an exhibit illustrating the whole process of wood-engraving and printing in China, Korea, Japan, and Tibet, also an exhibit of writing-materials, paper, brushes, inks, pigments, and ink-pallets. The oldest printing blocks in existence are likewise preserved in the Museum. They are engraved with floral designs and must have been made before the year AD 1108. They were found in the ancient city of Chu-lu in southern Chili, which was submerged by a flood in that year.

The results obtained by the economic process of block-printing are stupendous. It is best adapted to the genius of Chinese writing which employs many thousands of characters, and has many advantages over movable types which are expensive, difficult to store and to arrange and hard to find when needed in setting type. Block-printing could easily be established anywhere and made literature accessible to everyone at a moderate cost; it is a democratic art. Above all, it has always satisfied the aesthetic sense of the Chinese in that the block-prints preserve accurately the beauty of form of the characters and the hand-writing of the individual. No two Chinese hand-writings are alike, and hardly two block-prints can be found in the same style of writing if based on the hands of different individuals. In typography, of course, the type is standardized and conventional.

One of the disadvantages of block-printing was the storing, arrangement, and preservation of blocks which were easily destroyed by humidity or fire. For the printing of the *Tripiṭaka*, the sacred scriptures of the Buddhists, a copy of which is in the Newberry Library (printed in

1730-38), 28,411 blocks were required.[16] It was customary, especially under the Ming, in the government printing office at Nanking, if single blocks were lost, to re-engrave these and to add the date on the margin. Thus, the Newberry Library has several editions of the dynastic histories made up from blocks of different dates such as 1368, 1530, 1531, 1533, 1572, etc.

As to the form of the Chinese book, it was originally in the form of a roll. The manuscripts written on silk under the Han were kept in rolls, likewise all manuscripts and xylographs from the second down to the tenth century. The Buddhists introduced the folded book, somewhat on the order of our railroad time-tables, and still retain it for their sacred literature. The stitched and paged book, as we have it now, is not older than the Sung dynasty and goes back to the eleventh century. How it originated is not yet ascertained.

The earliest printed book in existence in America is preserved in the Newberry Library, Chicago, and is dated AD 1167. It is entitled *T'ang Liu sien-sheng wen tsi* and contains, in twelve volumes, the collected poems and essays of Liu Tsung-yüan (AD 773-819), one of the celebrated poets and essayists of the T'ang dnasty. The work next in date secured by me for the Newberry Library is a general history of China known as *Tung kien kang mu* by the philosopher Chu Hi, published in 1172; and it is a complete

---

[16] The original woodblocks for the printing of a much earlier edition of the *Tripiṭaka* is preserved in the Haeinsa Monastery, South Korea. Carved in the 13th century, the set consists of a total of 81,258 woodblocks. This represents the earliest and most complete extant version of the Buddhist canon in Chinese characters.

copy of this first edition in a hundred volumes, which is in the possession of the Library. It is a rare and fine specimen of Sung printing and perhaps the most extensive work of that period now known. Several Sung editions are also in the Library of Congress and in the Gest Chinese Research Library at McGill University, Montreal. One of the recent acquisitions of the latter is an edition of the *Tripiṭaka* of the Sung and Yüan periods, of different dates, the earliest actual print going back to AD 1232. This collection was first printed in AD 972.

The Chinese were also the first who conceived the idea of the printed newspaper. The Peking Gazette, "the News of the Capital" (*Ching Pao*), is the oldest daily paper in existence. It began to appear in AD 713 under the T'ang dynasty, and has since been issued daily until the collapse of the Manchu dynasty in 1911. It contained the imperial rescripts, decrees, and all official news relating to interior and foreign affairs. It was printed in two editions, in a Government edition sent to officials throughout the empire, and in a popular edition sent to regular subcribers in the capital and the provinces; also a manuscript edition could be obtained.

One of the forerunners of printing is represented by inscriptions carved in stone tablets. China is a land of inscriptions, many thousands of which are still in existence, and epigraphy is a favorite occupation of her scholars. As early as AD 175 the texts of the Confucian classics were inscribed on stone for their permanent preservation. Subsequently these texts were deeply incised in stones, and paper rubbings were taken from them for distribution among scholars. Stone tablets were the recognized method of preserving exact copies of fine calligraphy and drawings of great masters. Taking rubbings of all sorts of inscriptions

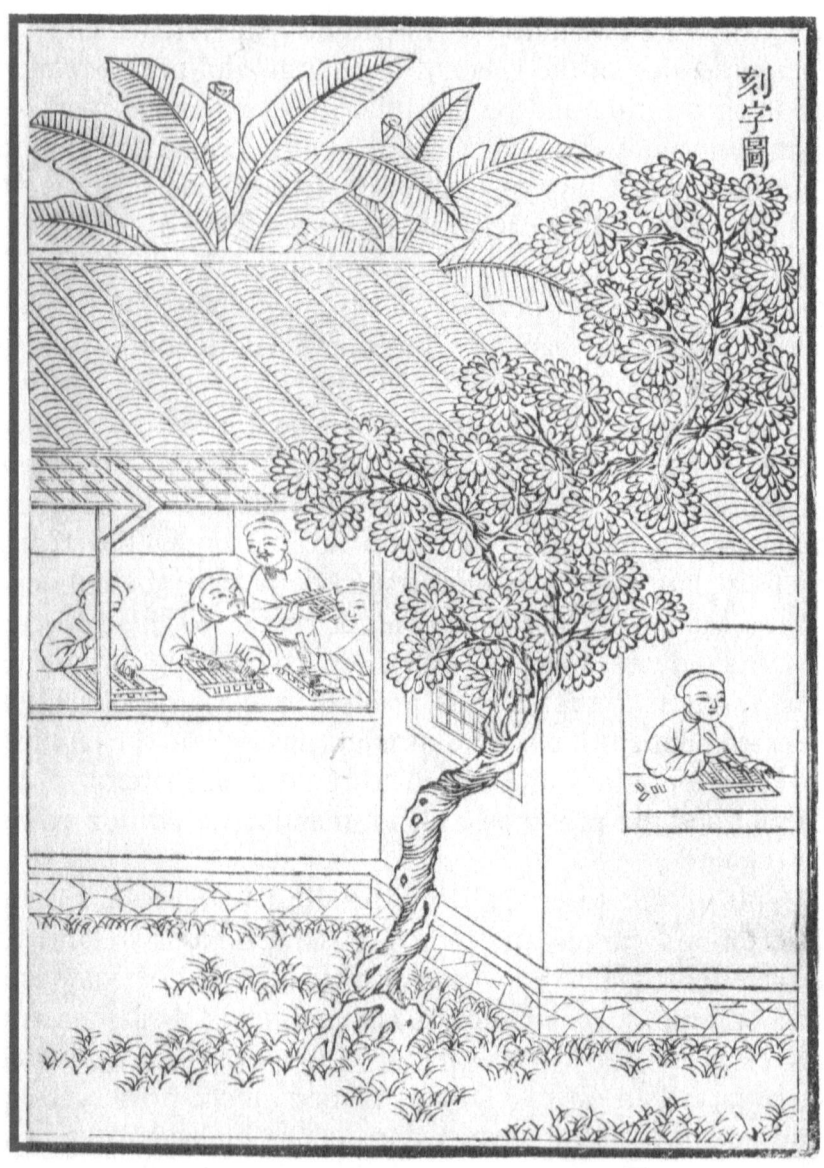

Fig. 9. Cutting movable Chinese type, from an old xylographic print reproduced in *A Chinese Printing Manual* [Manual for Wu Ying Palace Movable Type, 1776]; privately printed for the Zamorano Club, LA 1954.

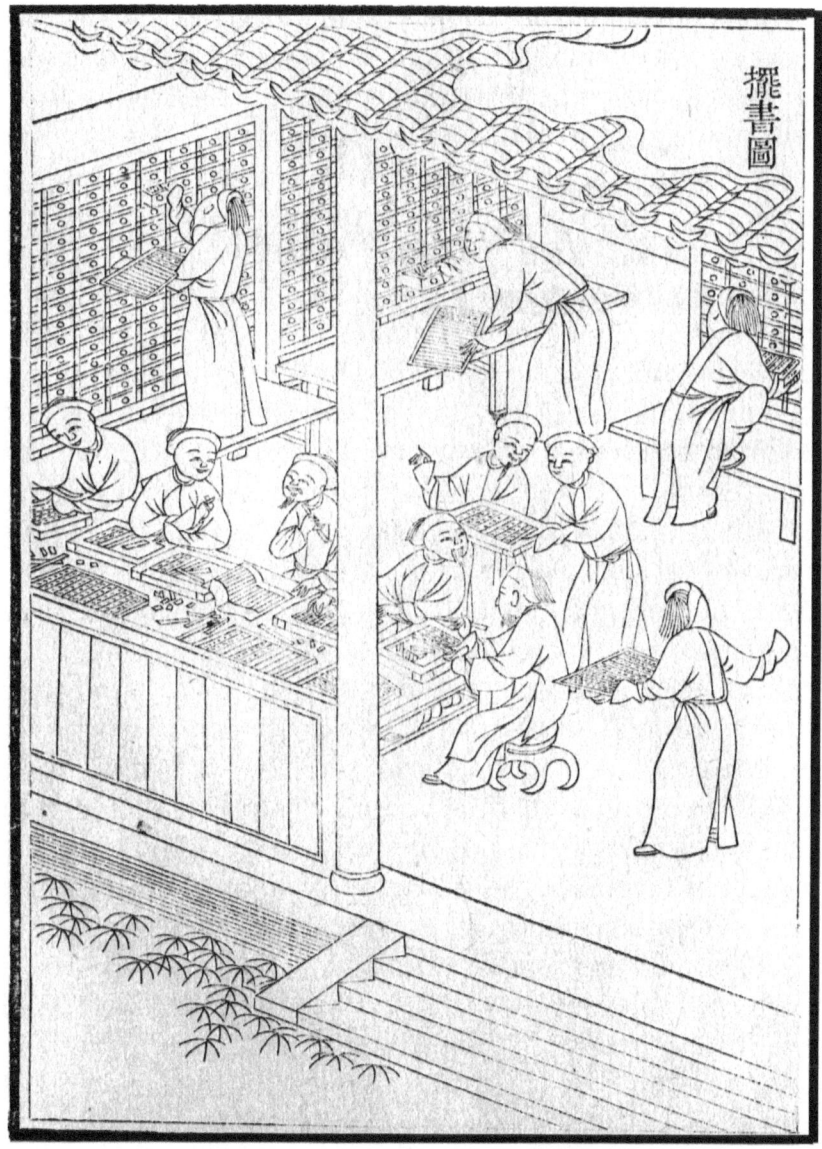

Fig. 10. Setting movable Chinese type, from an old xylographic print reproduced in *A Chinese Printing Manual* [Manual for Wu Ying Palace Movable Type, 1776]; privately printed for the Zamorano Club, LA 1954.

has developed into a regular trade since the days of the T'ang dynasty, and in Chinese collections are still preserved rubbings taken in the Sung and Ming periods, chiefly from inscriptions now lost, that are highly prized.[17]

To obtain facsimile rubbings of inscriptions on stone or bronze, the Chinese use sheets of thin, but tough paper, which is folded, slightly soaked in water, and then evenly applied to the surface of the inscription. The paper is pressed in with a wooden mallet and forced into every depression by means of a soft brush. When the paper becomes sufficiently dry, they go over it with a stuffed pad of cotton lightly dipped into liquid ink. When taken off the paper shows a perfect inpression of the inscription coming out in white on a black background. The men doing this work form a special profession, and as ancient inscriptions are numerous, there is a lively trade carried on in such rubbings to supply the demands of scholars.

At an early date the Chinese experimented with movable type. In the period AD 1041-49 a commoner, Pi Sheng by name, is said to have made a set of clay types which were locked in an iron frame for printing, but no print made from these types has survived. Under the Mongol dynasty, in the fourteenth century, type was cast of tin and subsequently made of wood. Wooden types were made by Wang Cheng, a geographer and agriculturist, who likewise devised a revolving table upon which the types were arranged, and from these wooden types he printed his *Nung shu*, a work on agriculture. In a chapter of this work published in AD 1313 he records a history of his set of movable types, stating that

---

[17] Figure 5, Ch. I, is an example of a Song dynasty rubbing of a prized ink-stone, bearing an inscription, that has survived to present times, while the ink-stone itself has long been lost.

the characters were first engraved in wooden blocks which were then sawed apart into individual types.

Printing from movable metal type on a large scale was first practiced in Korea,[18] and reached there its highest development. A set of 100,000 copper types was cast in AD 1403 by order of the king, and actually used in the publication of many books up to the year 1544. The Japanese Government General of Chosen reported in 1916 that it had taken care of old types of metal, clay, and wood, formerly in the possession of the imperial household of Korea, to the extent of about half a million pieces.

A revival of type printing took place in China at the end of the seventeenth century. At the suggestion of Jesuit missionaries, the emperor K'ang-hsi had a font of 250,000 movable copper types cast which were used for the printing of an extensive cyclopaedia, the *T'u shu tsi ch'eng*, in six thousand volumes, completed in AD 1726. In AD 1736 there was a shortage of currency, so that this font was sent to the melting pot for the minting of copper coins. It was replaced in AD 1773 by a font of wooden type which was used for printing the catalogue of the emperor Ch'ien-lung's library and other books. Printing from movable type was an expensive undertaking requiring large capital and was entirely carried on by the government, ceasing when government support

---

[18] The oldest surviving book printed with movable metal type dates to 1377, the *Jikji Simche Yojeol* ('Anthology of *Seon* [Zen] Teachings of Great Buddhist Sages'). It was written by a Korean *Seon* Buddhist monk in Chinese characters and printed in Korea, and is now held in the Bibliothèque Nationale, Paris. There is additional archaeological evidence of the use of movable metal type in Korea that dates back perhaps a century earlier.

was withdrawn. Block printing was found more practical and reasonable for private and commercial purposes. By the nineteenth century the use of type had come to an end in China, Korea and Japan, and was reintroduced from the West as an entirely new art. At present European typography and even paper and printer's ink dominate China, but one cannot say that the production of these modern presses are as elegant, graceful, and artistic as the time-honored block-print.

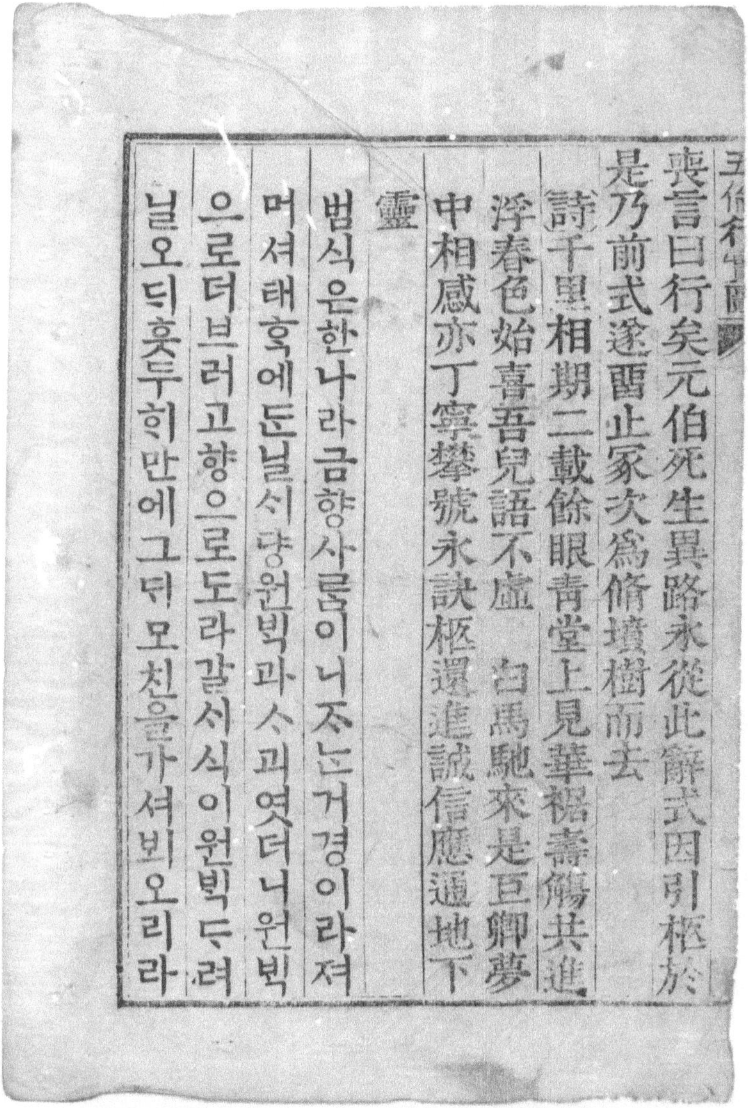

喪言曰行矣元伯死生異路永從此辭式因引柩於
是乃前式遂酉止家次為脩墳樹而去
詩千里相期二載餘眼青堂上見華裾
浮春色始喜吾兒語不虛白馬馳來是巨卿夢
中相感亦丁寧攀號永訣柩還進誠信應遍地下
靈
범식은한나라금향사롬이니조는거경이라져
머셔태혹에둔닐시댱원빅과소괴엿더니원빅
으로더브러고향으로도라갈시식이원빅드려
닐오듸훗두히만에그딕모친을가셔뵈오리라

Fig. 11. Sample page of book printed in Korean Chongnija movable type, a copper-alloy font cast in 1795. The book, typeset in both Chinese and native Korean (Hangul) scripts, was printed on mulberry paper in 1797. Private collection.

INDEX

Page number entries in italic indicate illustrations.

A
America, first paper mill, 82
Ancient Khotan, 21, 62
Annals of Japan, 56
Annals of the Later Han, 7, 78
Annam, Chinese ink in, 21
Antiquities of India, 68
Arabs, and rag paper, 1; introduction of paper to, 80; paper introduced in India by, 65, 86
Asia, Central, history of ink in, 60ff

B
Ball, J. Dyer, 36
bamboo, component of paper, 10, 81, 83; writing on strips of, 3-7, 9, 12, 76, 77
Barnett, L. D., 68
Beckmann, 67
birch-bark, mss written on, 65
block-printing—*See* woodblock printing
Blümner, 67, 68
bone, as component of ink, 30, 50; as conveyance of writing, 4; use in divination, 75
book, Chinese form of, 92
Book of Historical Records, 5
Book of Rites, 5
Book of Songs, 5
Bower manuscript, 65
Brahmanic literature, colored inks in, 69
brush, writing, 2, 3, 5, 9, 10
Buddhist clergy, as pioneers of block-printing, 87

Bühler, G., 65, 68

C
Cambodia, documents written on deer skin, 71; paper and ink introduced from China, 71
Carletti, Francesco, 51
Castrén, A., 63
Cave of the Thousand Buddhas, 22
Champion, Paul, 45, 55
Chang, Jen-hi, 19
Chang, Kü-tsing, 28
Chang, Yü, 27, 33
Ch'ao, Kien-tse, 7
Ch'ao, Shwo-Chi, 15, 16, 17, 27
charms, as objects of first mass printing, 87
Ch'eng, Kün-fang, 40
Ch'eng, Lao-po, 17
Ch'eng shi mo yüan, 40, *43*
Ch'ien-lung, 61, 97
China—See Davis, John Francis
China, reverence for written word in, 64, 86; writing materials in the earliest antiquity of, 3; 73
China grass—*See* ramie
Chinese Commercial Guide, 48
China, Chinese, and ink, 2, 3; attempts made by France to imitate Chinese ink, 45; book-making, 40; culture, 3; ink in *materia medica*, 50; ink in Turkestan, 62; ink superior to that of Japan, 56; inventions, 1; literature on ink, 17; prominent qualities of Chinese

ink, 44; rivalry with Europe, 45; technology, 3—*See also* dynasties (China)
Ching Pao, 93
Chi p'u, 18
Cho keng lu, 12, 18
Chou li, 5
Chou Ta-kwan, 71
"Christian Art in China," 40
Chu Fung, 27
Chu shu ki nien [Annals Written on Bamboo Tablets], 3
coal, bituminous, used for writing, 13, 15
colored inks, 37, 60, 69, 70
Contancin, P., 38
cords, knotted for conveyance of messages, 4, 73
Couvreur, Séraphin 9
cuttle-fish, 36

D
Damascus paper, 82
Diashi, Shōtoku, 56
Davis, John Francis, 36
De Guignes, 49, 51
De Mély, F., 14
Description de la Chine, 25, 37, 45
Dictionary of the Economic Products of India, 69
"Discourse on Ink," 19
divination, 4, 9, 75, 88
dragon fragrance compound, 33
Du Halde, 25, 37, 45, 46, 48, 51
Dunhuang—*See* Tun-huang
dynasties (China): Chou, 4, 5, 7, 76, 77; Han, 3, 6, 11, 12, 14, 20, 44, 54, 76, 78; Kin, 18; Kitan, 30; Liang, 52; Manchu, 19, 51, 93; Ming, 18, 19, 25, 32, 37, 39, 44; Shang, 4; Sui, 13, 21; Sung, 18, 19, 25, 28, 32, 39, 44, 89, 92, 92, 96; T'ang, 16, 17, 18, 21, 22, 24, 25, 26, 27, 28, 39, 42, 44, 54, 89, 92, 93, 96; T'si, 17; Tsin, 2, 3, 7, 12, 14, 15, 18, 20; Wei, 2, 3, 7, 12, 15, 16, 18, 20; Yüan, 12, 17, 18, 44, 93

E
Egypt, familiarity with ink, 2; paper in, 82; opposition to printing, 86
Entwurf einer Beschreibung der Chines, 19
Europe, medieval, familiarity with ink, 2; opposition to printing, 86
"eyebrow-paint ink", 27

F
Fabricius, B., 67
Fan Ch'eng-ta, 35
Fang I-chi, 28
Fang shi mo p'u, 39, *41*
Fang, Yü-lu, 39
Faraj, Abu'l, 38
Feldhaus, F. M., 2
Feng Tao, 89,
Field Museum of Natural History, 30, 49, 54, 61, 62, 91
Four Departments of the Study, The, 18
France, attempts to imitate China ink, 45
Fung Chi, 17, 44
Fryer, John, 66

G
Geerts, 2, 14
Geschichte der Erfindungen, 67
Gest Library, 93
gifts, ink as favorite, 42
Giles, 2

Glossary of Reference on Subjects connected with the Far East, 2
gold, writing, 60-61
Goshkewich, I., 19
Gough, A. E., 69
graphite, 14, 15, 78
Gray, L. H., 66
Gutenberg, 72

H
Haeinsa Monastery—*See Tripiṭaka*
Handbook, State, of the Chou Dynasty—*See Chou li*
Han Dynasty, Later, annals of, 7
Han Hi-tsai, 27
Havret, H., 48
Hoernle, 65
Horiuji Monastry, 66
Ho Yüan, 18
Hua mei mo—*See* eyebrow paint ink
Huang-ti, 3
Hüan Tsang, pilgrim, 65
Hüan Tsung, Emperor, 25, 42
Huichou factories, 46-47
Hui Tsung, Emperor, 35
Hu King-shun, 32

I
Imperial Ink, 25, 33
India, cotton stuffs and birch bark used for writing letters, 65; learning in, 64; palm leaf manuscripts in, 66; paper and printing in, 65; wooden writing boards in use in, 65
India ink, 2, 36, 66, 67, 68
Indische Palaeographie, 65
Industries anciennes et modernes de l'empire chinois, 45, 55
Industries of Japan, 2
ink balls, 20

ink-cakes, 13, 19, 24, 26, 26, 30, 39, 40, *41*, 42, *43*, 59
ink factories, early, 24, 25
ink-fish, 36—*See also* cuttle-fish
"Ink History," 18
"Ink Memoirs," 18
"Ink official," 24, 26
"Ink of nine children," 39, *41*
ink-pallets, 18, 53, 54, 91
ink, Chinese, 21; colored, 60, 61, 69; oil combustion, 33; gold, 60, 61; preserving, 44; shell, 20; silver, 60, 61; vermilion, 61
ink-stones, *53*—*See also* ink-pallets
ink-writing, oldest extant specimen of, 65
inventions, Chinese, 1
ivory, tablets, 4, 76

J
jade, tablets, 4, 76
Jametel, M., 2, 19
Japan, Chinese ink in, 21; Horiuji Monastery, 66; imitates Chinese and Korean ink models, 58; ink imported into China from, 49; ink inferior to Chinese, 56; manufactures ink for home consumption, 56; printing of charms in, 87; process of ink manufacture, 56-58
Japanese Homes and Their Surroundings, 59
John Crerar Library, 40
Julien, S., 2

K
K'ang-hsi, 61, 97
Kanjur, 61
Kao-li—*See* Korea
Kao Lien, 19

Kao Tsung, 25-26,
Kia Se-hie, 15, 16, 17
Ki kung, 20
Kitan, ink of, 18
Ko-bai-en bokufu, 42
Korea, conveys ink manufacture to Japan, 30; first to manufacture ink from lampblack, 28; improves on paper-making and printing, 30; ink-cakes, 30, 32; ink-cakes offered as sacrifices to the gods, 30; ink not for medical use, 34; Korean ink, 19, 29; printing from movable types cast of copper, 30; tribute of ink to the court of China, 27-28, 34
K'ou Tsung-shi, 34
Ku mei yüan mo p'u ("Collection of the Inks of the Old Plum-tree Garden"), 42
Kün-fang—*See* Ch'eng Kün-fang
Kwang chou ki, 14

L

Lamas, Mongol, 63
lampblack, 12, 13, 15, 16, 27, 28, 32ff; first used in manufacture of ink, 19; fundamental substance of ink, 54
Lao-tse, 73
La Provence du Ngan-hoei, 48
leather, documents inscribed on, 62
Le Compte, Louis, 35, 45
L'Encre de Chine, 2
Le lapidaire Chinois, 14
Les documents chinois découverts dans les sables du Turkestan oriental, 11
Les produits de la nature japonaise et chinoise, 2, 14
Li Fang, 17

Li ki—*See* Book of Rites
Li Kung-lin, 45
Li Lung-mien, 45
Ling wai tai ta, 35
Li Shi-chen, 8, 13, 16
Lu Ki (Lu Shi-heng), 13
Lung hiang tsi—*See* dragon fragrence compound
Lun mo—*See* "Discourse on Ink"
Lu Yu, 17, 18, 62

M

Mahābhārata, 64
Manchu ink, 63
Man t'ang mo p'in, 19
marking ink, 69, 70
Ma San-heng, 19
Masson, Charles, 65
materia medica, ink in, 34, 50
medicine, ink as, 20—*See also materia medica*
Memoirs and Observations made in a Late Journey through the Empire of China, 45
Mi Fu on Ink-stones, 53
mo (ink), 6
Mo chi, 19
money, paper—*See* paper
Mo ki—*See* "Ink Memoirs"
Mo king, 16, 17, 18, 27
Mongolia, writing in, 61, ink, 63
Mong-tse, 8
Mong k'i pi t'an, 13, 34
Mo p'u, 18
Morse, Edward S., 59
Mo shi, 17, 18, 62—*See also* "Ink History"
Mo tsien, 19
movable types—*See* types, movable
mulberry bark, 80, 81
Mundy, Peter, 37

INDEX

Mung T'ien, 9, 10
Mu t'ien tse chwan, 6

N
Nan-yü—*See* Ting Yün-p'eng
Newberry Library, 42, 91, 92
Nichols, Henry W., 49
Nung shu, 96

O
oil-combustion ink, 33
oldest recipe for preparation of ink, 15
Old Time Wall Papers, 85
orpiment, ink made from, 37

P
pagoda-paper—*See* wallpaper
Palladius, 2
palm-leaf, ink for, 69
palm-leaf manuscripts, Buddhist, 66, 70
P'an Heng, 30
P'an Ku, 28, 30
paper, as money, 82; early existence, 3; from bamboo, 81, *83*; from bark, 11, 78, 79, 80, 81, 82; from bast-fiber, 10, 11, 70, 78; from hemp, 11, 78, 79, 81; from rags, 1, 2, 10, 11, 77, 78, 79, 80, 85; from silk refuse, 10, 11, 77, 78, 80; from silkworm coccoons, 10, 30, 77; sized, 82; treatise on, 18; wrapping, 82; writing, 82—*See also* wallpaper
Papers rel. to the Collection and Preservation of the Records of Ancient Sanskrit Literature in India, 69
papyrus, 65; superceded by paper, 82
Parker, E. H., 20

Peking Gazette—*See* Ching Pao
Pelliot, Paul, 22
Pen ts'ao kang mu, 8, 13, 19, 51
Pen ts'ao yen i, 34
Periplus of the Erythrean Sea, 66, 67
Persia, paper factories in, 82; use of Indian ink in, 66
petroleum ink, 34
pi (stylus), 5, 7
Pie lu, 14
Portuguese, first printing press in India set up by, 65, 87
preserving ink, 44
printing-blocks, earliest use of, 87, 88; oldest extant, 91; preparation of, 48, 90; preservation of, 91, 92; printing with, 22, 48, 49; smoothing with ink, 26
printing, first became known, 21; oldest specimen of, 22, 88; opposition of elites to, 64, 86
proverbial sayings, ink in, 1, 52
punishment, liquid ink as a, 52
P'u Ta-shao, 33

R
Ragionamenti sopra le cose da lui vedute ne'suoi vraggi, 51
rag-paper, 1, 2, 10, 77, 80, 85
ramie, 80
recipe for making ink, Chinese 15, 17, 37-38; Japanese, 57
Rein, J. J., 2
Ricci, Matteo, 40
rubbing, *53*, 93; technique of, 96

S
sacrificial ink, 30
Schott, W., 19
seals, as a prototype of printing, 86
Se-ma T'sien, 3, 9

Serindia, 21, 22
shell ink, 20
Shen Kwa, 13
Shi king—*See* Book of Songs
shi mo (stone ink), 12
Shōchou, 26, 27
Shön Kwa, 34
Shön Kwei, 32
Shōtoku Daishi—*See* Daishi, Shōtoku
Shōtoku, Empress, 87
Shwo wen, 8
Siam, Chinese ink in, 70, Buddhist scriptures written on palm leaves, 70
Siao Tse-liang, 17
silk, refuse, paper made from, 10, 11, 77, 78, 80; tablets, 11
silkworm cocoons, paper made from, 10, 30, 77
silver, writing in, 60-61
Sino-Iranica, 35, 67, 70
Srong-btsan sgam-po, 60
Stein, Sir Aurel M., 11, 21, 22, 62, 88
Subandhu, 66
Süe t'ang mo p'in, 19
Su I-kien, 18
Sung Lao, 19
Sung Ying-sing, 19

T
T'ai p'ing yü lan, 17
tallies, wooden, 4, 73, 75
Tan-cheng, 56
T'ang Liu sien-sheng wen tsi, 92
Tanjur, 61
T'ao Hung-king, 36
T'ao Tsung-i, 12, 18
Ta-yo—*See* Ch'eng Kün-fang
Technik der Vorzeit, 2
Technologie, 67

Ten Bamboo Studio, letter paper from, 72fn
Thailand—*See* Siam
Things Chinese, 36
Tibet, Chinese ink in, 21; colored inks in, 60; culture of, 60; ink powder, 49; printing books in, 48; use of writing slate in, 61;writing introduced from India, 60
Tien Chen, 3
T'ien kung k'ai wu, 19
Ting Yün-p'eng (Nan-yü), 39, 40; design of, *41*
tortoise-shell, as conveyance of prophetic inscriptions, 4
treatises on ink, 12, 15, 17-19
"tribute ink," 27
Tripiṭaka, in Gest Library, 93; in Newberry Library, 91; woodblocks preserved in Haeinsa Monastery, 92fn
Ts'ao Chi, 19
Ts'ao Ts'ao, 13
Ts'ai Lun, 11, 73, 78, 79, 80
Tschichold, Jan, 72fn
ts'i (varnish), 5
Ts'ien Fu Tung—*See* Cave of the Thousand Buddhas
T'si min yao shu, 15, 17
Ts'in Shi, 9, 36, 77
Tsu Min, 24
T'u Lung, 19
T'ung ya, 28
Tun-huang, 88
Turkestan, Chinese, documents inscribed on leather, 62; ink of, 18, 21, 44, 62, 63; rag-paper preserved in, 80; wooden tablets recovered from, 11, 12, 76
T'u shu tsi ch'eng, 17, 97
Tu Wan (Yün-lin), 14

two colors, inks of, 20
types, movable, Chinese use of *94, 95*, 96, 97; Korean use of, 30fn, 97, 99; material used for, 30, 96, 97; oldest extant example of, 97fn
typography, Chinese, as fine art, 72; compared with woodblock printing, 91; of Europe, dominance in China, 98

U
Up the Yang-tse, 20

V
Vajracchedikā, 22, *23*, 88
Vāsavadattā, 66
Vietnam—*See* Annam
Voyages à Peking, 1784-1801, 49

W
wallpaper, 84, 85
Wan, Duke, 5
Wang Hi-chi, 1
Wang Kiün-te, 16, 24
Wang Chie, 88
Watt, G., 69
Wei lio, 28
Wei Tan (Wei Chung-tsiang), 16, 17, 19

Wen fang se p'u—*See* "Four Departments of the Study"
Williams, S. Wells, 48, 49
wine-lees, 67
woodcuts, 18, 19; oldest extant example of, 22, 88
wood, writing on, 4, 5, 6, 11, 12
woodblock printing, advantages over movable type, 91; earliest use of, 21, 87, 88; oldest extant example of, 22; popularization of, 89; technique, 48, 89-91
workmen, in ink, status of, 25
Works of the Russian Mission of Peking, 19, 20
writing-brush, 1, 2, 3, 9, 10, 51, 59, 77
writing materials, in earliest antiquity of China, 3
wu-tse fish—*See* cuttle-fish

X
Xinjiang—*See* Turkestan, Chinese
xylography—*See* woodblock printing

Y
Yen Shi-ku, 9
Yi shwi kung mo—*See* "tribute ink"
Yün lin shi p'u, 13
Yün sien tsa ki, 17, 44

ABOUT THE AUTHOR

Dr. Berthold Laufer (1874-1934) was one of the most eminent and accomplished sinologists of the 20th century. Born in Cologne, he received his doctorate from the University of Leipzig in 1897, having, in the process, developed fluency in Chinese, Japanese, Manchu, Mongolian, Tibetan, Russian and most other European languages.

Recruited to work for American institutions by anthropologist Franz Boas, Laufer undertook three long and gruelling expeditions to East Asia (the Russian Far East, Japan, China and Tibet) between 1898 and 1911. He returned with a vast collection of artefacts and manuscripts from these regions, providing the basis of the rich Asian collections of the Newberry and Crerar Libraries, the American Museum of Natural History, New York and the Field Museum, Chicago, the latter becoming Laufer's employer for the remainder of his career.

As curator of anthropology at the Field Museum, Laufer enjoyed considerable status within that institution and financial stability for his family; nevertheless he found the intellectual environment of life in Chicago to be stultifying. His solution was to immerse himself in his scholarship and writing. Over the course of his career, terminated by his untimely death in 1934, Laufer wrote over 450 monographs, books and reviews, an accomplishment all the more impressive when taking into consideration the incredibly broad scope of his studies. Laufer's publications encompassed not only Asian anthropology, but also the ethnology, archaeology, art, languages and religions of that vast region, even detouring at times into such areas as diverse as the history of plant cultivation and animal husbandry to name but two. Many of these works remain standard references on the subject addressed. His accomplishments cemented his reputation as both a scholarly maverick—in the best sense of the word—and the most influential sinologist of his generation. Laufer was, in the words of a recent Field Museum retrospective,[19] "the greatest in his field then, and one that has had few equals since".

---

[19] Quoted from 'Curators. Collections and Contexts: Anthropology at the Field Museum, 1893-2002,' *Fieldiana – Anthropology*, New Series, No. 36, September 2003. For further details on Laufer's remarkable life and career, see this publication, as well as "Berthold Laufer, 1874-1934," *Journal of the American Oriental Society*, Vol. 54, No. 4, December, 1934. The latter includes an extensive bibliography of Laufer's publications.

www.ingramcontent.com/pod-product-compliance
Lightning Source LLC
Chambersburg PA
CBHW031431210526
45464CB00005B/2156